SAND STONE
AND SEA STACKS

SAND STONE AND SEA STACKS

A Beachcomber's Guide to Britain's Coastal Geology

RONALD TURNBULL

F

FRANCES LINCOLN LIMITED
PUBLISHERS

Frances Lincoln Limited
4 Torriano Mews
Torriano Avenue
London NW5 2RZ
www.franceslincoln.com

A catalogue record for this book is available from the
British Library.

978-0-7112-3228-0

Printed and bound in China

1 2 3 4 5 6 7 8 9

PAGE 1 Ammonite fragment, Saltwick, near Whitby. The outer coils
are a reproduction, in siltstone, of the original shape. The inner coils
are a cast: an impression left on the siltstone when the original shape
dissolved away. The fossil has been mineralised in pyrites.

PAGE 2–3 Mountain Limestone stack, Three Cliffs Bay, Gower.
The seawashed beach rocks here show fossil coral and shells.

RIGHT Youths of Portstewart, Antrim, defy death off the dolerite.

CONTENTS

INTRODUCTION: SEASHELLS AND THE OCEAN OF TRUTH

On the granite of Land's End, looking to the 'Armed Knight' seastack. Wolf Rock, beyond, is not granite. Mountain building also involves volcanoes. The Alpine mountain building left just one UK volcano, its plug being Wolf Rock.

INTRODUCTION: SEASHELLS AND THE OCEAN OF TRUTH

I do not know what I may appear to the world, but to myself I seem to have been only like a boy playing on the sea-shore, and diverting myself in now and then finding a smoother pebble or a prettier shell than ordinary, whilst the great ocean of truth lay all undiscovered before me.

Isaac Newton (apocryphal, widely quoted)

William Smith, the founding father of English geology, was to mock Newton, an inhabitant of the Oolitic Limestone, for not looking at those pebbles hard enough …

Newton's own fields, or at least those he must have often walked over, are literally strewed with fossils in a manner which I never saw in any other soil, lying thereon like new-sown seeds of oats, and so numerous are they where I observed them, that in the moist state of that tenacious soil the great philosopher may have scraped them (unobserved) from his shoes by hundreds. It was this which, on my receiving the Wollaston Prize, induced me to say that 'had Newton condescended to look on the ground he must have been a geologist'.

William Smith, 1839, quoted in Phillips, 1844

We do like to be beside the seaside – rock pools and paddling, sandwiches with real sand in them, striped towels and a big yellow umbrella to keep off the sea breezes and summer showers. It's a peculiarly British form of fun, which foreigners might even misinterpret as suffering. But after greasing yourself with a mixture of suncream and grit, cleaning beach oil off your feet, and abandoning the attempt to read the latest Dan Brown with a Force Six breeze turning the pages … look above and around you. What Britain is and where it came from, just what's been going on around here for the last 500 million years: all is revealed in a continuous slice around our seaside.

Here are giant tree ferns engulfed in volcanic lava. Here are caves, rock-arches and sea stacks; here are desert sand dunes emerging out of the ocean. Here are cliffs bent and crumpled by two continents crashing into each other; and bands of red-hot rock squeezed out by a volcano somewhere up in Scotland. And as you wander along the edge of the sand, gradually slowing your eye to the beach-holiday speed of looking at things, you see small creatures, seashells and corals from hundreds of millions of years ago. All along the Dorset coastline are curled ammonites as if someone tried to unravel the workings of the great pocket watch of Time and guess what – the little springs inside it came pinging out across the rocks.

But take a poet to the seaside and he tends not to have a jolly time. Matthew Arnold visited Dover Beach in autumn, with a big sea crashing up the flint shingle: its 'melancholy, long, withdrawing roar' reminded him what a dreadful state England was in. (He didn't spot the fossil sea urchins.) For his visit,

Fast Castle, on the Berwickshire coast of Scotland. It stands on a tough, ocean-floor sandstone called greywacke, the default rock of the Southern Uplands.

ABOVE Wave-cut platform south of Stonehaven, where the sea has also chewed out caves and a sea stack. The rock is conglomerate – sandstone with rounded cobblestones in it – of the Old Red Sandstone.

BELOW Old Red Sandstone at Cove Harbour

RIGHT The author sits at Siccar Point on the celebrated Hutton's Unconformity. Below, tough greywacke sandstones have been upended, then cut off above by an invading ocean of 300 million years ago. On top lies red sandstone laid down later on the bed of that same ocean. In the base of the red sandstone are lumps of greywacke, part of the shingle beach that lay on top of the ancient wave-cut platform.

TS Eliot wore grey flannel trousers with their bottoms rolled, and failed to hear the mermaids calling. Thomas Hardy, on the Cornish coast, looked across the dark spaces of the sea and thought of a woman lost forty years before. The waves on the cold grey shores of the Isle of Wight made Tennyson too think about death.

We come to the edge of our busy land, so worked over with footpaths and fenceposts and bits of litter, and gaze over a vast emptiness. A couple of masonry corners stand against the sea; a fringe of stonework tops off 60 metres of vertical greywacke rock. I came down to Fast Castle on the coast of Berwickshire as the last sunlight was a brownish glow across the dead heather. The breeze whistled through the dry grass, and far below, the sea was whispering amongst the rock stacks. A seagull soared to castle level, let out a cackle, and dropped back into the shadows.

A metre of land connects the castle with the moorland above: there's a path of crumbling concrete protected by an old chain. Crossing it, you look down lichened beds of tough grey-black sandstone: beds so bent by earth movements that they stand on edge, straight out of the sea in smooth slabs that take an evil gleam under the sunset. Southeast, a single lighthouse, and a rock tower like a second castle against the fading sea. The other way, I looked up the coast to the glow of Torness nuclear power station, and two black lumps

against the sunset, the volcanic plugs of Bass Rock and North Berwick Law.

Fast Castle must have been quite something, when it was still there. It was here that Princess Margaret Tudor, after 13 years among the elegant gardens and tapestried halls of Hampton Court, spent her first night in Scotland on her way north to marry King James. I imagine her descending the rough heather on her pony to where black towers stood against the darkening sea, and a lantern above the gatehouse, and the smell of the salt water. Fast Castle is now where not only Margaret Queen of Scots but also this book's author has spent the night. The facilities are not what they were in 1503 but then, Queen Margaret didn't have a Gore-tex bivvy bag or a plastic box of Co-op luxury coleslaw.

Gore-tex bivvy bags are slippery, and I didn't fancy a midnight descent onto the black sea stacks 50 metres below. So I retreated to the moorland at the mainland end of the iron chain, and bedded down in the deep heather. All night long, the sea stretched away towards Denmark – reminding me, as it must have reminded Queen Margaret in 1503, that most of our planet is actually under water. While the seacliff steepness, dropping to darkness and a flash of breakers and the harsh cry of sea birds, tells how the sea inexorably moves inwards, battering the little solid bit that we scamper about on.

The sea moves forward against the land, like a very slow bulldozer with a blade as wide as the North Sea. The breaker crashes, and the withdrawing water rubs at the rock with the handful of sand it carries. In this game of water versus rock, the sea has time on its side. It carves a cave; the cave becomes an arch, and then a sea stack; and the sea stack falls into the sea. The backwash carries away the rubble, and the wave blade moves forward another metre.

The wave-cut platform at Robin Hood's Bay, on the North Yorkshire coast, has been sliced out by the sea in the 20,000 years since the Ice Age. That's an average of 2 cm per year: the speed of a small child growing. Yet that would be enough, in 5 million years, to meet the sea on the other side. In geological terms, 5 million years is a mere moment. We Johnnies-come-lately, human beings, have been around almost as long as that. The planet has been here long enough for the sea totally to destroy the UK a thousand times over.

In Norfolk the sea is cruising inland much faster. Elsewhere, the waves are obstructing their own project by piling shingle against the cliff base. Changes in ocean level can send the sea back to the start to cut a new platform 5m lower down. So Britain may not disappear exactly as described above.

But then again, it already did. Twice. The Greensand Sea chopped off the top of southern England, from the Weald across to the Devon coast. And it happened before that at Siccar Point, a few miles north of where I lay watching the sunset fade above Fast Castle. It's a place made famous by a man called James Hutton, Scotland's first geologist.

James Hutton's insight about the world came in a letter to the engineer Joseph Black: 'The world? The world is no

more than a turnip!' This seems to have been a catchphrase of his, as one sardonic friend described Hutton and Black as 'two philosophers rioting over a boiled turnip'.

And Hutton was a man who knew his turnips. A so-called 'gentleman farmer' in a very ungentlemanly land, 'a cursed country where one has to shape everything out of a block and to block everything out of a rock'. Over the rough country of Berwickshire, he carved drains, struggled with boulders, and introduced the lighter Suffolk plough. His innovative crop rotation included turnips, which were drilled in straight lines to allow hoeing between the rows – so the turnip crop cleared his weeds before going on to feed his sheep. He also introduced the practice – still common – of undersowing the barley crop with cultivated grasses: harvest the barley and there's your hay field all ready to go.

James Hutton looked at the world as he looked at his turnips. Looking at the turnips, he saw the sense of sowing them in straight lines. Looking at the land below them, he saw rivers carving through the rocks, carrying away sand into the sea. He saw the sea bashing away the cliff bottoms of Berwickshire: 'In the natural operations of the world, the land is perishing continually.' And he saw all of that sand and debris re-forming, at the sea bottom, into the next lot of rocks. 'As the present continents are formed from the waste (mineral but also animal and plant) of more ancient land, so, from the destruction of them, future continents may be destined to arise.'

Rocks are made out of earlier sand: sand is made out of earlier rocks. It's a process apparently endless and without any visible beginning.

Cockburnspath, end of the Southern Upland Way and start of my walk to Fast Castle, has a stone cross at the village centre, a small shop, and a handsome sandstone church hidden among the cottages. At Cove Harbour I passed over round-edged, red-layered sea cliffs, rocks that were obviously made, as Hutton realised, out of earlier sand. The Old Red Sandstone was laid down in rivers or shallow lakes, or still preserves the swirl of a desert sand dune.

The wide beach of Pease Bay was the last chance to stop for a swim; and many were doing so, as the tide moved in over the sun-warmed sand. Behind the bay, though, not layered sea-cliffs, but caravans, laid in parallel lines like the natural rock strata that they so conspicuously were not. To travel across the country in a wheeled metal box, so as to live for a few days inside a larger metal box? Many find it a satisfying form of fun, going by the size and number of caravan sites all around Britain's coastline. (And a caravan really is a whole lot better than a rucksack for carrying those interesting beach pebbles home in.)

Beyond Pease Bay the Southern Upland Way turns aside, the lanes end, and the cliffs rise in hard grey rocks to rough fields where turnips grow out of red-brown sandstone soil. At the field corner is an interpretation board, and a steep-dropping grass slope (there's a fence to hang onto) down to the sea and Siccar Point. It was in 1788 that James Hutton arrived here in a small boat, along with his young colleague John Playfair, and a Mr Hall whose father owned the boat. And here, where the red sandstone lies down against the ancient, tilted, grey rocks of Fast Castle, he and those seashore rocks displayed to his friends the incredible ancientness of the Earth.

First the tough grey rocks, themselves a sort of sandstone, had formed in some deep ocean. Then they had been tilted sideways, and raised to form dry land. Some time – a very long time – after that, the sea had flattened the dark grey continent and overrun it. Red sand had covered it, a grain at a time, hundreds of metres deep; and been crushed underneath another ocean into a different, reddish sort of sandstone.

And after all that, the land rose again into the air. Once again, the sea attacked it. And the old junction of the grey and the red re-emerged, excavated by the waves, to be seen by Mr Hutton and his friend Mr Playfair in their little boat. But before reckoning up all that amazing span of time, just think of one more thing. That grey rock, right at the start, is itself a sort of sandstone. It formed at the bottom of the sea from small grains brought down by rivers from some even earlier land … Playfair, brought up with the idea of a Biblical Earth just 5,700 years old, was stunned.

We felt ourselves necessarily carried back to the time when the schistus [the tough greywacke sandstone] on which we stood was yet at the bottom of the sea, and when the [newer, red] sandstone before us was only beginning to be deposited, in the shape of sand or mud, from the waters of a superincumbent ocean. An epoch still more remote presented itself, when even the most ancient of these rocks, instead of standing upright in vertical beds, lay in horizontal planes at the bottom of the sea, and was not yet disturbed by that immeasurable force which has burst asunder the solid pavement of the globe ... The mind seemed to grow giddy by looking so far into the abyss of time.

Geologists seeing Siccar Point for the first time are sometimes so excited they burst into tears.

So at teatime I stopped off at Siccar Point to pay respects to the spirit of James Hutton – and found a TV crew already at it. They'd arrived the authentic way, by sea, and were filming the famous red-meets-grey unconformity in the last moments before the cliff shadow fell across the rocks. They'll be geologising Scotland this winter, unless somebody at BBC 2 happens to think up a new six-part drama about nurses or policemen.

You can fit a lot into this six-mile stretch of Scottish coastline. The abyss of time in the afternoon: overnight, the near-infinite ocean.

At dawn I headed up from Fast Castle to the signposts of the new extension of Berwickshire's Coastal Path. They pointed me along field edges and past a decomposing combine harvester, intricate girderwork under the early pink rays of the sun. It wouldn't have looked out of place as an artwork in the Turbine Hall of Tate Modern, and fitted equally well at the corner of its clifftop field. But then the path swung around a gorse bush to the tops of great grey cliffs. Those cliffs dropped 150 metres into a sea that was slaty green, and fringed with white foam in the early sun. A stream chuckled under a footbridge, fell over the cliff edge, and was snatched by the easterly breeze in circles around the next big curve of the cliff.

LEFT Siccar Point, on the North Sea coast of the Scottish Borders. Just south of Hutton's Unconformity is a wave-cut platform of today into the Old Red Sandstone. The further cliffs are of the older and tougher greywacke.

RIGHT Siccar Point: a tough bed of greywacke that the ancient sea could not subdue pokes up through the overlying red sandstone. In the background, the same greywacke has been carved into exactly the same formation by the ocean of today.

FAR RIGHT Siccar Point: the ancient sea plucked out a vertical gap in the underlying greywacke; the gap was then refilled with red sandstone sediments.

And here the rocks changed again, from ancient grey to orange and mauve, the volcanic lava of St Abbs Head. On the purple rocks there was pink thrift in great bright bunches, and yellow birdsfoot trefoil. Not even at the Tate Modern – only Nature herself is brave enough for such a colour scheme. Out of green seawater rose the orange and mauve sea stacks, but the ledges of them were dull grey with thousands of squatting guillemots.

The cliff path looks down across tottering chimney-stacks of red basalt to St Abbs village – after the night in the wilderness, a civilised cup of coffee for breakfast.

The two mighty voices

Two Voices are there – one is of the Sea,
One of the Mountains; each a mighty Voice ...

So wrote Wordsworth, in a poem suggesting that Switzerland and England were the two Top Countries when it came to resisting Napoleon. Proper geologists poke around among the dumped fridges and rusty barbed wire of old quarries, or the litter-sprinkled cuttings of our road system. But when it comes to looking at rocks in places that are actually worth looking at, we take the hint from William Wordsworth; we go to the seaside – or up into the mountains.

My earlier book *Granite & Grit* was the mountain one. Of the eleven geological periods, it covered the first six, and the last: it so happens that the Permian, Triassic, Jurassic, and Cretaceous are not represented among UK mountains. The present book is the seaside sequel. And it happens that Britain's seaside resorts offer a full survey of the second half of our island's rocky story, from the Old Red Sandstone right up to the last few million years.

Up in the mountains you see ice and fire: the granite from the roots of ancient volcanoes, the gneiss and schist twisted

by earth movements of 400 million years, the carving of the glaciers. Down at the seaside you see – obviously – the sea. The sea is, after the glacier, the other great shaping agent of our land, as it is at this particular (geologically unimportant) moment. But you also, as it chances, see the sedimentary rocks, the ones themselves formed by the much more ancient seas: the sandstones, limestones and shales.

The mountain rocks are mostly old; they also tend to be igneous (newly congealed out of molten magma) or else metamorphic (squashed out of all recognition). Either way, they tend not to offer fossils. The UK seacoast, once you know where to look, is rich in the fossils of ancient seashells and coral.

Back in the eighteenth century, a 'fossil' meant any interesting object your drain-digger man dug out of your estate or your nephew brought back from the Empire; just provided it was small enough to fit in your Chippendale display cabinet. A couple of Roman coins; a set of 'St Cuthbert's Beads', small rosary-like stones from the beach at Holy Island; a travelling inkwell used by Bonnie Prince Charlie when he invaded Derbyshire; a flint shaped like someone's penis. Proper geologists out of Oxford and Cambridge sneered at such oddments as hopelessly unscientific – at least until they realised your Cuthbert beads were a surprisingly useful set of Carboniferous crinoid segments. And then – the beasts! – they bashed open the interesting flint in case it had a sea urchin in it.

The museum at Whitby is still arranged in that confusing, intriguing, eighteenth-century way. The mummified hand of a murderer; a Japanese iron dragonfly; a world-class collection of fossil plesiosaurs. And so is this book.

The proper way to cover 8,000 miles of coastline, over 550 million years, is by a series of tough but instructive remarks along the lines of: the Glamorgan coast of South Wales shows strongly folded strata of Dinantian age unconformably underlying a typical Lower Jurassic facies, with a well-displayed *Gryphaea* and ammonite fauna. As it happens, that strongly folded Dinantian has been voted Britain's best beach. You'll find a photo of it, along with the unconformity and the *Gryphaea* (or Devil's Toenails) in Chapter 8. But this does mean that many, many other equally well-displayed places are skipped over or ignored.

In particular, I've ignored the Scottish Highlands – including what *should* have been voted Britain's best beach, Sandwood Bay in Sutherland. The rocks of northern Scotland are different from, and older than, everything else in the UK. They are beautifully displayed at Sandwood Bay and other places. The beaches of Wester Ross, though beautiful, are chilly in climate and infested with midges. And their rocks are seen even better in the mountains overhead. So in this book, there's very little of northern Scotland, and not much of the Precambrian, Cambrian, Ordovician and Silurian periods either.

I do like to be beside the seaside. But even better, I do like to *see* beside the seaside; the sea stacks and wave-cut platforms, the ammonites and the sand, longshore drift and continental collisions, rubble washed out of a mountain range that – hang on, where's it gone? It was right here a moment ago. Between the lichen and the low-tide line, everything's out in the open to be looked at. In the foreground, limestone alternates with shale and a very small coal measure. And at the other end of the bay, the Great Whin Sill marches out into the sea.

Various commercial firms will polish up your geological samples to reveal their structure and texture. Or the Bristol Channel will do the job for free. The sea even obligingly brings down bits of the cliff so as to scatter fossils across the shoreline.

So paddle in the rock pools, or lie around on your stripy towel, picking up an interesting pebble. Grab a Cornish pasty and stroll along the cliffs. Real geology isn't reading up the books and cleverly memorising all the long words. Real geology is looking at real rock, and working out what the heck's been happening to it.

LEFT Basalt sea stack, St Abbs Head, Berwickshire

BELOW White Rocks, County Antrim. The sculpture park display of sea stacks and arches shows the standard Antrim colour scheme – white chalk below, black basalt on top – which will prove useful in Chapter 3 for identifying volcanic vents and in Chapter 7 for tracing faultlines. At the very base of the wave-cut platform is a glimpse of the brown sandstone and shale underlying the chalk. It will be crucial in solving the 'Case of the Impossible Ammonites' in Chapter 3. Beyond, the north winds have formed Ireland's second-largest sand dune. Curran Strand is convex, an unusual shape for a beach: it's caused by waves diffracting inwards from either side of the offshore islands seen to right. Those islands, the Skerries, are the tilted-up edge of a big dolerite sill. Not all the images in this book will contain such an exhausting amount of rock interest!

1. THE SEA

Lias cliffs and wave-cut platform, Traeth Bach and Mawr, Glamorganshire. The Bristol Channel has twice the tidal rise and fall; so it can potentially carve out twice as much wave-cut platform.

1. THE SEA

The sun had not yet risen. The sea was indistinguishable from the sky, except that the sea was slightly creased as if a cloth had wrinkles in it. Gradually as the sky whitened a dark line lay on the horizon dividing the sea from the sky and the grey cloth became barred with thick strokes moving, one after another, beneath the surface, following each other, pursuing each other, perpetually. As they neared the shore each bar rose, heaped itself, broke and swept a thin veil of white water across the sand. The wave paused, and then drew out again, sighing like a sleeper whose breath comes and goes unconsciously.

Opening words of *The Waves* by Virginia Woolf (1931)

Just two of Earth's minerals occur in liquid form. Mercury makes silver-mirror droplets in rock cavities around the occasional volcanic vent. Water, by contrast, lies kilometres deep over most of the planet.

Other places in the Solar System don't have oceans. Mars once had flowing water, but there's none there now. Venus has water, enough to form a surface 'ocean' 20 cm deep, but all of it is vapour and cloud in the greenhouse-heated atmosphere. Hyperion, a small moon of Saturn, has an ocean of solid ice that may be 50 km deep. Everywhere else is pretty much dry.

Some places are too small to have sea. Any ocean on Mercury would evaporate into space, the individual water molecules moving fast enough to escape that planet's weak gravitational field. The lightest gas molecules, hydrogen and helium, can similarly escape the stronger gravitational field here on Earth. Water molecules, fortunately for us, are too heavy to get away.

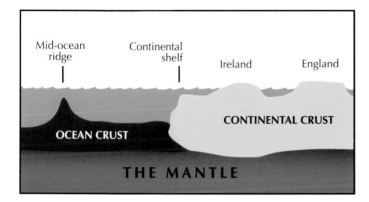

LEFT In the battle between water and rock, water always wins in the end. Sandwood Bay, just south of Cape Wrath – the sea stack, in Torridonian sandstone, is Am Buachaille.

ABOVE AND RIGHT The coastline of today is arbitrary, depending on sea level and sea erosion. The continental shelf is the true edge of Europe and the underwater 'shoreline' of the Atlantic. When we come to track back the opening of the Atlantic in Chapter 5, it's this continental shelf that will match up with the continental shelf of America.

The comets that lurk in the darkness at the edge of the solar system are made of dirty ice. Comets striking Earth soon after it first formed may have given us our oceans. However, careful weighing of our water tends to disprove this one. There are two types of hydrogen: the normal, and the double-weight deuterium that forms 'heavy water'. Once mixed, these two sorts are difficult to separate out again: one reason why H-bombs are tricky to make. The ratio of the two hydrogens is quite different in the comets from what it is in Earth's oceans. Earth's water, then, is likely to be from somewhere else: we didn't get wet as a result of a cosmic snowball fight.

Today there's a new theory. Soon after the Earth congealed out of the dust cloud around the young sun, something hit us. A mini proto-planet crashed into our forming surface and melted huge areas of it. The remains of the mini-planet, along with some melted bits of Earth, splashed away to make what is now the Moon. And Earth was left with an atmosphere of red-hot melted rock, along with all the gas and water vapour dissolved in those rocks. As that rock-based water vapour cooled, it rained and rained for thousands of years, filling up the oceans.

What difference does it make, all this water sloshing around? The ocean may be essential for the processes of plate tectonics – the way our planet's surface shifts around and recycles itself. In order for oceanic crust to be pulled by its own cold weight back into the earth's mantle, water is the lubricant. Above the descending ocean plate, continental granite-type rocks melt, and bubble upwards, bringing fresh minerals towards the surface. Here again, water is the flux, the solvent chemical that helps the granite melt. The circulation of the ocean bed carries wet ocean-bottom sludge back

into the Mantle, at the same time as the Mantle delivers fresh basalt and water vapour at the mid-ocean ridge.

Stand on a cliff top and you realise that ours is a water planet, interrupted by episodes of land. As space-based telescopes get bigger, we shall be discovering whether blue planets like our one happen all over the place, or not. The only life we know so far occurs and persists on the only wet, splashy planet we know; which is also the only planet we know about with plate tectonics. This may mean something, or it may not. We're the only life we know: but in cosmic terms, we don't know much.

All life is sea life, if you consider our bloodstream as an internalised ocean. But up until the Devonian period, 400 million years ago, life was sea life in the more normal sense of living in the sea. Much of it still is. Not, on the whole, in the deep ocean, where the useful minerals needed to build bodies are about 3,000 metres below the essential sunlight needed to sustain them (or the plantlife food itself sustained by sunlight). But 150 miles west of Ireland, the ocean floor steps upwards by 2,000 metres. This is the continental shelf. At the top of it, we're in a different sort of sea. Not only is it much shallower: we're now swimming above continental crust rather than the darker, heavier oceanic crust. We're on rocks that were once land, and that will be land again next time there's an ice age to lower the ocean, or a continental collision to raise the rocks.

The continental shelf is a better place to live. And one of the liveliest parts of the continental-shelf sea is where its edge meets the land. Rocks and sand give a stable home life to seaweeds, as well as to shellfish and worms which live by sieving seawater for tiny plankton. Higher lifeforms like fish and crabs head in to the land-sea boundary for the freshly washed-out minerals, as well as for delicious human sewage and the warm water from nuclear power stations.

And from the other direction, the lifeform that considers itself the highest of all flocks down to the land's edge, looking for some sand to lie about on.

The cruel sea

The sea is what makes our planet; but it's also what unmakes it again. We spread our beach towels on the finely broken debris of what was once our landscape. The cliffs on either side are the wound left by the slicing of the sea. You can take all your ice ages and volcanoes: the real natural disaster is the one you roll up your trouser legs and go paddling in, any time you stroll down the beach.

The sea is not so much a bulldozer as a strimmer, with string that's a mere 2 or 3 metres thick. The wave action slices, precisely, along the layer between the low tide line and the high. The surging sand and pebbles act as a giant grinder. The sea seeks out weaknesses in the base of the cliff, carves out a notch, then carves out a cave. The cave falls; the backwash carries away the rubble.

The result is a wave-cut platform: a rock shelf exposed at low water, covered up at high tide. At its inland end, the sea-strimmer attacks the cliff base. If the rocks happen to be tougher at the top than the bottom, as they are along the Yorkshire coastline, then the cliff, demolished from its base, gets undermined and overhangs.

Thus the base of the cliff will stand at, or just below, the high tide line. The tide sneaks in across the flat platform to the base of the overhanging cliffs; and innocent rock pool paddlers and geologists get trapped against the cliff foot. The wave-cut platform is Nature's way of getting rid of geologists.

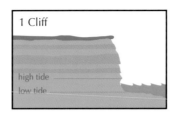

1 Cliff

high tide
low tide

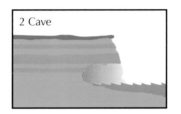

2 Cave

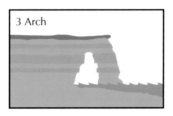

3 Arch

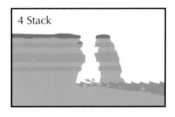

4 Stack

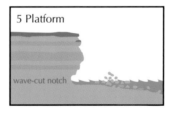

5 Platform

wave-cut notch

Most cliffs have cracks: faultlines from some long-ago movement of the earth. The sea gets into the cracks, hauls away the broken rocks, and probes deeper. Where a faultline happens to run right through a headland, a sea arch will be the result. Given the way the sea attacks only at the bottom, arches are almost inevitable – which doesn't stop them from being a wonderful inversion, rock resting on top of air. And which doesn't stop unimaginative inhabitants from naming the feature, as often as not, the 'Needle's Eye'.

Sooner or later, though, stone must cease to be suspended over air. The arch falls, and its outward end becomes a sea stack. The sea gnaws around its base, and after a century or two the sea stack falls into the sea. The broken stones are carried away by the waves, and the sea starts work on a new sea cave slightly further inland.

Life's a beach

You walk the cliff top, with fields and fences and little woods stretching inland to streetlit towns and motorways. The other way, just a big blue nothing, with waves crashing far below. On your left, civilisation: on your right – the sea. The trees are thorns, half-dead and growing sideways instead of up. Sea-mist dampens the yellow grasses. You feel small and unimportant and not particularly cheerful. But then the path turns down a ferny hollow. Stone steps, fuzzy with grey lichen, lead to a lane. The lane descends, 1 in 4 and a grinding of

LEFT Given 10 million years or so, the sea turns Britain back into sand and gravel. The slightly tiresome bit, from the sea's standpoint, is breaking down and clearing the loose rubble. A slowly rising sea level, as at the end of an ice age, makes the whole thing easier.

BOTTOM LEFT Sea cave on Ailsa Craig

BELOW When cliffs are of chalk or limestone, the sea water doesn't just batter them, but also dissolves them away. Stacks and arches are at their most sculptural when carved from chalk. White Rocks, Antrim

gearboxes, between arching Fuchsia bushes. At the bottom there's a stone jetty and an old iron winch and a bucket-and-spade shop. Each white cottage has an orange pantile roof, and a green fishing float in the garden.

Life's a beach: but is it a sandy one, or one of stones?

Common sand is small crystals of quartz, and quartz is important. Silicon and oxygen are two of the lighter elements in Earth's mixture. In the slow stirring of the moving, recycling tectonic plates, they end up like the pips in marmalade: at the top. The simple combination of two oxygen molecules with one silicon is silica; and silica is the mineral that makes continental rocks different from the dense, heavy, black oceanic crust. Most of the landforming minerals – feldspar, mica – are other elements combined with silica. It's silica that gives continental rocks their greyish and whitish colouring. It's silica that determines the behaviour of continental-type volcanoes (and that behaviour is bad behaviour, as we'll see in Chapter 4).

TOP This sea stack is just one of the fine sea features formed by a single limestone bed, the Portland Stone, along the Dorset coast. It will in due course be undermined, topple into the waves, and be carried away as beach shingle. Two others to its right have already suffered this fate.

RIGHT Just occasionally, a sea cave breaks out not into a sea cave on the other side but upwards, at the top of the cliff. On stormy nights the waves funnel upwards, and spray shoots out of the clifftop with a hollow roar. Well, that's what it does in Victorian novels anyway. Blowhole at White Rocks, Antrim.

The crystalline form of silica is quartz. Quartz is silicon-plus-two-oxygen in a three-dimensional array that's literally rock solid. Among common minerals, quartz is the toughest. Granite rocks break down in the wind and the rain; the mica and the feldspar decay to clay; the quartz crystals endure. The clay and quartz get compressed into sandstone; the sandstone weathers away; the quartz crystals again endure. They swish around in the sea, and the sea dumps them in quiet corners along the shoreline. Well, the corners are quiet until someone builds a car park and starts renting beach umbrellas ...

Quartz is colourless, and a pure quartz beach is gleaming white. But a little iron in the sand turns it golden yellow, the way beaches ought to be.

Pounded sea-shell makes a beautiful beach, creamy white. But shell is calcite. Calcite is fairly tough stuff, but it's a lot softer than quartz; also, it dissolves in sea water. So shelly beaches are rare. Limestone cliffs break down to limestone pebbles, and those again are calcite. Limestone lands, unless with a sand supply from somewhere else, tend to be beach-free.

Black basalt rocks, with sufficient sea-pounding, can make black sand. Somehow black sand doesn't quite do it. The volcanic Canary holiday islands are forced to import proper-coloured sand from the Sahara.

There are also black volcanic beaches on the Isle of Skye, and seashell beaches in the Hebrides. But in general, where geology hasn't supplied a nice pile of ground-down quartz, we spread our beach towels on shingle. On a sandy beach, the backwash of the wave carries sand seawards, making a smooth gentle slope. On shingle, the wave seeps down between the stones with a faint distinctive hiss. There's no backwash; and shingle beaches are steeper.

On a sandy beach, you can carry your shrimp net for ever down to the low tide line. A shingle beach, being steep, is consequently narrow. Not only is your towel spread on uncomfortable lumps – there's less beach to spread it on. The waves are steep, and sudden, against the shelving shore. They even sound different. Dover Beach in Matthew Arnold's poem has its 'melancholy, long, withdrawing roar' just because it is in quartz-free chalk country, and made of flint shingle. If Arnold had only been at Dover Beach, Barbados, which is a sandy one, he would have written an upbeat holiday

ABOVE Find some sand and sit on it. Beach life at Eccles, East Anglia

RIGHT (FROM TOP LEFT) Black sand at San Sebastian, Tenerife; Shingle beach below Alnmouth golf course; Shelly beach, Mhealasta, Outer Hebrides; Sea cabbage, *Crambe maritima*

brochure, rather than one of the most dismal poems in the English language.

Geologically speaking, there's something to be said for the shingle. When you've seen one tiny quartz crystal, you've seen them all. But shingle is interesting, sea-wetted at eye level, as you crawl on hands and knees, up the steep pebble bank at Budleigh Salterton. The waves even manage to arrange a varied selection, by a trick called longshore drift.

All along England's south coast, the prevailing wind is southwesterly, at an angle to the shoreline. When a wave hits the beach, it carries sand and pebbles upslope, but also slightly to the east. The backwash returns it straight down the slope. Without the wooden barriers called groynes, all the sand on the beach would eventually migrate off its eastern end. At the same time, liver-coloured quartzite pebbles from Budleigh Salterton, and crystal rhyolite ones from Dawlish, migrate by millimetres along the 18 miles of Chesil Beach, and occasionally end up at Brighton.

Walking the Cleveland clifftops, you notice quite fresh-looking fences wandering out into empty air above the sea. Farmers and regular coast walkers are well aware of how the cliff edge is constantly moving inwards. In 1993, a guest looked out of a bedroom window of the Holbeck Hall Hotel, south of Scarborough, and noticed that 50 metres of the hotel garden had gone. Over the next two days, watched on live TV, the four-star hotel disappeared over the cliff edge. Heavy rain had turned the glacier clay to mud, and this slid to the beach to bring the cliff edge back by 70 m.

LEFT Pwllstrodur, north Pembrokeshire. Storm waves fling sand, gravel and larger cobbles up the beach. The weaker backwash carries the gravel and sand down again, but leaves behind the big cobbles. Sand ends up at the bottom of the beach, cobbles at the top.

ABOVE Storm beach at Porlock Bay, east of Exmoor. Once in ten years, or fifty, a big storm sends cobblestones further up the beach slope than any ordinary storm can wash them back down again. Meanwhile sea water seeps through and creates a grass-killing pool on the inland side.

The Holderness coast south of Flamborough Head is entirely made of boulder clay left behind by the glaciers, and is eroding away at an average 2 m per year. This is faster than any other shoreline in Europe. The villagers of Happisburgh in North Norfolk may be surprised to learn that fact. In the last half century, all of the eastern edge of Happisburgh has fallen into the North Sea. But Yorkshire tykes should be nervous too. At the current rate of cliff crumble, a mere 50,000 years will leave Yorks smaller than either Cumbria or Devon – both those counties having older and tougher rock edges.

The sea takes it away. Just sometimes, the sea gives some of it back. The 18 miles of Chesil Beach have been donated by the sea out of debris from all the cliffs to the west. Longshore drift piles the shingle in a long strip butted up against Portland Bill. The sea even sorts the shingle into sizes: pea-sized at West Bay to mangoes and grapefruit at the Portland end. Even in thick mist, a barefoot fisherman could tell where he'd landed by the size of the stones.

The Yorkshire coast has its back to the prevailing south-westerlies, so it's the occasional storms from the north that shift the broken bits of cliff. Material eroded from the Holderness coast moves gradually southwards, to form the three-mile shingle spit at Spurn Head. Celebrated on TV in 2005 as one of the wonders of Yorkshire, Spurn is bleak but not strikingly beautiful. Its mud, sand and shingle are held together with coarse grasses, sea holly, and the sea buckthorn that appears so weird in autumn with its thin yellow leaves and orange berries. The single sandy road runs to a disused lighthouse with a black water tank perched on top. The remains of Victorian sea defences stick out of the sand like sculptures in timber and iron. The soundtrack is soaring curlews, and the harsh honks of the Brent geese. The spit is a motorway stopoff for migrating birds in various directions, so it's a place for spotting rarities. Along with the sea grass, the other common form of wildlife is the bird person.

Spurn Head might eventually become an island and re-attach to the southern side, thus migrating from Yorkshire into Lincolnshire. Two medieval villages, Ravenspurn and Ravenser Odd, were unwisely built on the shingle. Both of

ABOVE This sea loch isn't a fjord only because it's in Scotland not Norway. Loch Hourn, north-west Highlands

them have disappeared as the spit marches two metres westwards each year into the mouth of the Humber.

A large number of British beaches are utterly uncomfortable to spread your towel on. So it's surprising that Natural England considers the shingle beach an endangered resource. Shingle beaches grow hardly anything at all, due to the shifting around of the stones; so anything that does grow is going to be strange and special. As well as Natural England, builders of nuclear power stations like shingle, not just because it isn't covered with sunbathers, but because of the steeper underwater shape, and because the constant addition and removal of stones leads overall to a stable situation. And so Dungeness Power Station in Kent overlooks Derek Jarman's shingle garden, with its driftwood and spiky sea cabbage.

Ice, the sidekick

In the mountains, glaciers are the great land-shifters. Mountain ridges and corries, glens and gouged-out lakes are as the ice left them. At the seaside, we mostly see the sea – and especially so on England's south coast, where the glaciers never reached.

But even at the seaside, not all of the scenery has been shaped by salty water. Water in its frozen form reaches sea level as great glaciers. These gouge out the ground way below the level of the sea before giving up and letting themselves float away as icebergs. The glacier melts away, and we are left with steep-sided and astonishingly deep sea inlets – Loch Eriboll, on the north coast of Sutherland, is 100m deep. Elsewhere they are known as fjords; but being British, we call them sea lochs.

Where glaciers didn't gouge out sea inlets but ran out of steam on dry land, they left whole valleys filled with ground-up ground: mud, powdered rock and large stones, in a mixture called boulder clay. Boulder clay makes an unexciting sort of sea cliff of mud and stones, gradually washing away into the sea. That boulder clay is our youngest and

uppermost rock layer, and it's one we seldom see inland because of the plants growing out of it.

Among those plants are the turnip-sized onions and
football-sized turnips of the suburban allotments of England.
Most of the clay so cursed by people trying to grow pota-
toes is boulder clay from glaciers. But gardeners ought to
be grateful to the glaciers. Clay soils hold in the moisture.
And down at the tips of the monster parsnips and carrots,
the ground-up rocks gradually give out their minerals. One
reason for the dust bowl of the Southern US in the 1930s
was that the farmers expected the soils to behave like the
clay-based soils of glaciated England – but they didn't.

Where the sea comes up against boulder clay, it washes
it away. And this frees up the stones formerly carried by the
glacier. Such stones can be from some quite surprising places
(though normally from the north).

At Sandsend Beach, the beach below the clay cliffs carries pebbles of yellow and crimson rocks that didn't come
from the volcanoes of Yorkshire, as Yorkshire never had any.

TOP LEFT Boulder clay, at Runswick Sands, brings strange stones from
the north.

TOP RIGHT Strangers on the shore: volcanic beach pebbles on
Sandsend Beach in sedimentary north Yorkshire

BELOW Not boulder clay, just a boulder. This glacial erratic (the word
means 'wanderer') was dumped on the Ayrshire shoreline by a glacier
out of the granite Galloway Highlands, 30 miles inland. Erratics of
granite are often easy to spot, their rounded shapes quite out of style
with the rocks they've landed up on.

I saw one such stone in a cabinet at Scarborough, and leant
forward to read the label. Where did they think it came from:
Scotland? Norway? The label said simply, 'Sandsend Beach'...
Even skilled geologists admit that a single stone from somewhere unknown is pretty much guesswork. Thus I can risk
suggesting that the pink-on-black granite looks to me like
the granite of Stonehaven, in northeast Scotland. Meanwhile
the yellowish syenite and the deep-underground one with
the black and red crystals look distinctly un-British, and

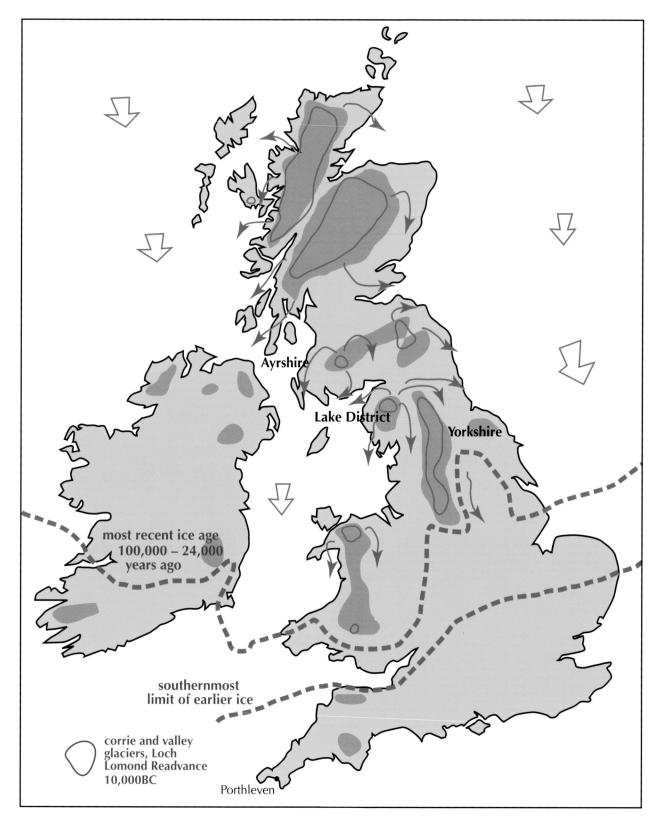

most recent ice age
100,000 – 24,000
years ago

southernmost
limit of earlier ice

corrie and valley
glaciers, Loch
Lomond Readvance
10,000BC

Ayrshire

Lake District

Yorkshire

Porthleven

ABOVE Recent ice cover in the UK. The blue-ringed areas are where permanent ice returned during the Loch Lomond Readvance, 12,000 years ago. These were also main centres of outflowing glaciers during earlier and icier times. Along the coast, there was constant pushing and shoving between major south-moving North Sea or Irish Sea ice, and local glaciers. Thus at different times the Yorkshire coast received stones from Norway, Scotland, and the Lake District.

If you find a strange stone south of the blue lines, then it wasn't brought by any glacier. At Porthleven, the 20-ton Giant's Rock, sitting just above the low tide line, is made of a garnet gneiss not found anywhere in the UK. So it must really have been dropped off by a giant.

make me wonder whether they've crossed the North Sea from Norway.

Ice melting in the Antarctic may, in the next decades, bring the sea level up a metre or more around our shoreline. But ice melting here in Britain, at the end of the Ice Age, has had the opposite effect. When ice lay on our island it weighed it down. Ever since the ice melted, 20,000 years ago, the land has been gradually bobbing back up again. Because snow arrived from the southwest, the ice cover was heaviest in the west. Some parts of the northwestern seaside have risen, since the ice left, by 200 m. More typically, the old beach from the Ice Age sea is now seen 10 m to 20 m above the current one. From the back of the beach of today rises a gently sloping field: a field well drained but hard to work, as its soil is ancient beach shingle. Sea stacks rise from the grassland, and at the back of the field is a former sea-cliff, complete with caves and even, for very lucky farmers, an elegant arch.

The sea takes apart the rocks, moving inland a thousand kilometres in a mere million years. But a juicy brown Yorkshire mudstone won't be kept down for ever. Powdered Yorkshire slides to the sea bottom, and gets compressed by more powdered Yorkshire on top. Mud and sand become, again, mudstone and sandstone. As continents collide and the world wrinkles up between, a new Yorkshire, mud-brown, sandy yellow, and full of fossils, rises into the mist and rain.

ABOVE Above today's beach and across the main road, a raised beach at Pinbain Bridge on the Ayrshire coast, with the former sea cliff and the sea stacks against which the ice floes once bobbed.

LEFT The Aradena Gorge, on the White Mountains coast of Crete, reveals not a raised beach, but a sunken gorge, and a spectacular sea-level rise. At various times in the last 50 million years the Straits of Gibraltar have closed up, and the Mediterranean has evaporated to a salty puddle. Dig away the sea shingle, and this freshwater gorge goes on downhill for another 200 m beneath the sea.

Saltwick Nab, on the Cleveland coast. Shales and mudstone below being attacked by the sea; sandstone standing firm at the cliff top; hence the steepness of Cleveland cliffs. Having converted mudstone back into beige mud, the wave backwash is carrying it out to sea. There it will drop to the bottom and, in due course, form some more mudstone.

2. UNDERSTANDING SAND

The Thirlstane, at Powillimont on the north Solway coast. The beach sand is stones and rock broken down by the sea. Meanwhile the cliff rock is sand, squeezed back together by the sea.

2. UNDERSTANDING SAND

It was Pythagoras who spotted that the planets progress across the starscape in patterns that, with a bit of mathematics, you can make some sort of sense of. But stones were just what you sat down on while you were looking at the stars.

And so they stayed for the next 1,500 years. Neptune (in charge of Earth Sciences as well as the sea) and subsequently the Christian God had built the ground out of various sorts of building stones so that we could build various sorts of buildings. Clever engineers knew how to find rocks that could be made to yield lead, copper, iron and gold. (A tip there: gold is caused by the rays of the sun: so look for it in the south, where those rays are at their strongest.) And in odd, uncouth parts of the world up near Hadrian's Wall they said there were black stones that burned – but anyone who'd been to Britain was bound to be a bit touched in the head.

It was the philosophers of the Renaissance who realised that it didn't matter what you were interested in: there would always be interesting stuff to be discovered. Drops of water had invisible living creatures swimming in them. Jupiter had moons. The insides of animals and condemned criminals had all sorts of stuff going on in them.

The first Renaissance man to see stones and think they might be interesting was a Dane called Nicholaus Steno. (Well, there was also Leonardo da Vinci. But Leonardo wrote it down in back-to-front writing in notebooks that nobody opened for several centuries.)

Steno was born in Denmark in 1638, the son of a prosperous goldsmith, but by the age of seven he'd lost both father and the subsequent stepfather. The Thirty Years' War ended

LEFT The sea, busy making sandstone and mudstone. Morecambe Bay from Humphrey Head, with Heysham Nuclear Power Station

ABOVE Not all sand is shaped by the sea. Windblown sand will give a differently textured sort of stone (seen in Chapter 10). Ireland's second-biggest sand dune, at Curran Strand near Portrush

RIGHT Ripples formed in sand by the last high tide (at Balcary Bay, Dumfriesshire), and approximately 400 million years earlier, in the Silurian (at Rockcliffe, a few miles to the east)

with these inspiringly honest words: "Instead of promising to satisfy your enquiring minds about the anatomy of the brain, I confess to you here, honestly and frankly, that I know nothing about it."

Grand Duke Ferdinando di Medici, ruler of Florence, was unusual among his family: he didn't flaunt his wealth and sophistication by being a patron of the arts. Instead, he did it by patronising the sciences. He adopted the brilliant youth into his court circle. In 1666, some Tuscan fishermen landed a great white shark. The Duke had its dead head hauled up to Florence to be anatomised, live in theatre, by Steno, his star scientist.

in 1648, but a plague six years later killed every third inhabitant of Copenhagen. In 1658, luckily still alive, he gained a place in the well-regarded University of Copenhagen, where an earlier graduate had been the great astronomer Tycho Brahe. At once the Swedes came raiding across the frozen ocean to besiege Copenhagen, and famine followed.

Denmark was Lutheran in religion, slightly apart from the wars and disasters of mainland Europe, and ruled by the science-supporting King Frederick III. It was a good home for any Renaissance scientist. Well, any scientist apart from one interested in the rocks – Denmark is flat (highest point 171 m not counting Greenland), and mostly made of sand.

Steno was better organised than Leonardo. His student notebooks covered medicine, alchemy, religion, mechanics, crystallography – occasionally even his official subject of study, medicine. Better organised than Leonardo's didn't mean much: he aptly labelled his journal 'Chaos'.

Continuing his studies in Amsterdam and Paris, Steno became one of the first media scientists, at a time when medicine was one of the performing arts. In the anatomy theatres of Europe's capitals he dissected, with brilliance and showmanship, anything put before him. He discovered Steno's Duct which links your salivary gland into your mouth. He found the Fallopian tube joining the ovary to the womb, thus establishing that woman is not just a passive recipient of seed from the man, but lays eggs internally like a fish. He established that muscles operate by contraction – they can pull but never push – and worked out the geometric linkages that still let them direct your limbs in all their different directions. In Paris in 1665, he started a lecture on the brain

LEFT Sun-cracked mud, Parrett estuary, north Somerset

RIGHT Mud cracks in sandstone, Crail Harbour beach, Fife

The year 1666 was when Newton fled Cambridge because of the plague, made fundamental discoveries in mathematics and optics, and formulated the law of gravitation. It is known as 'Annus Mirabilis', the wonderful year, not for that, and certainly not for Steno's shark adventure, but because of a successful sea-battle against the Dutch, and also because, despite its being numbered 666 for the Beast of the Apocalypse, its worst disaster was merely the Great Fire of London. The 304 verses of Dryden's poem don't mention Newton, let alone what Steno did with the shark.

Which was to notice that its teeth were shaped the same as strange objects called tongue stones from the island of Malta, which were used as a snakebite remedy, as he knew from his medical studies. This distracted him from medicine, into the science of geology – a science that was not just unnamed, but undiscovered. (Undiscovered, that is, assuming you couldn't read Leonardo reversed right-to-left.)

Three years later he published *Prodromus* (preamble) *to a dissertation on one solid naturally contained within another*. In other words, to a shark tooth or seashell contained within a piece of rock. He was the first to use the word 'stratum' to mean layers within the rocks, and formulated three laws of geology still fundamental today.

1. The principle of superposition: the stratum underneath is older than the stratum on top.
2. The principle of original horizontality: a stratum originates as a sediment laid down in water. Accordingly, whatever its current angle, it started off lying flat.
3. The principle of lateral continuity: sediments spread sideways. If you find a stratum in one place, you can expect to find it again elsewhere, or else find something specific that blocked it off.

Sandstone is made of sand

Sandstone is made of sand. That's obvious, and even more obvious if you try to clamber up sandstone seacliffs, which have a tendency to turn back to sand again under your flip-flops.

It's also obvious that the continents are solid and don't move around underneath us. It's obvious that volcanoes are on fire and spout out ashes. It's obvious that the sea can't carve out caves just with waves. Not everything that's obvious turns out to be true.

But some of it does. Steno was surprised that he just had to look, to see something so simple, and innocently wondered if there were other even more obvious facts staring him in the face. Probably nothing else quite so obvious – unless it's so obvious we still haven't seen it to this day. Galileo had already done the trick of dropping different sized stones off the Leaning Tower of Pisa. And a year before, in 1668, a Frenchman called François Petit had spotted the match-up of Africa and South America across the Atlantic – though the theory of Continental Drift would take another three centuries before it suddenly became as obvious as it is today.

Sand, and sandstone. The two pictures on page 35 show ripples formed in beach sand by the last high tide (at Balcary Bay, Dumfriesshire) and a few miles east and 400 million years earlier, in the Silurian period at Rockcliffe. This not only suggests that the stuff at Rockcliffe really is made of sand, it even indicates that the sand in question was under moving water.

The picture above left shows cracked mud on the foreshore of Parret estuary, north Somerset. Suppose someone dumps some building site spoil before the next high tide – yes, it is that sordid sort of foreshore. When the sea comes in, it washes coarse grit into those cracks. Wait around for 300 million years, and you get a rock that's something like the one found on the Fife coast, above right.

Sediments should be laid down in layers. But in the picture on page 32 of the Thirlstane, something strange is going on on the left, above the graffitist's letter R. The layers aren't flat, but slantwise. This is called cross-bedding. A stream flowing from right to left has been depositing layers of sand on the downstream end of a sandbank. (Cross-bedding, on a bigger scale, also happens on the downwind side of a sand dune.) On top of that, more sand has been laid in flat beds in still water.

Seen close up, the Thirlstane rock contains small quartz pebbles, like the ones in the Portishead sandstone on this page. Their rounded shapes show that they've been swept along in a stream. When a stream reaches the sea, the pebbles drop to the bottom. Coarse grit is carried slightly further, fine sand a bit further than that; a few miles out, just a thin silt drifts to the sea floor. These pebbles, up to 1 cm across, show that this sand was either in the riverbed itself, or if in the sea then very close to shore.

A channel sometimes fast flowing, sometimes still. Coarse gravel at, or close to, the shore. The suggestion is of a river delta, with its shifting channels. The geology book agrees with the suggestion.

LEFT Close-up of the pebbly layer at Portishead: water-worn quartz pebbles up to 2 cm across

RIGHT Worm-scrambled sandstone above Staithes: image shows 60 cm width of rock

BOTTOM LEFT Old Red Sandstone, south of Portishead

BOTTOM RIGHT Cowbar Nab at Staithes, Cleveland coast. The scrambled-up stuff at the bottom of the picture has been rearranged by worms and molluscs ('bioturbated').

How come this stone's got a seashell inside it?

> That is the case as it appears to the police, and improbable as it is, all other explanations are more improbable still.
>
> Arthur Conan Doyle,
> 'Silver Blaze' in *Memoirs of Sherlock Holmes* (1894)

When we look at sediments we see the pebbles graduating upwards like a sandbank; the fossil ripples on top; the cross-bedding; the stripes as tough sandstone interleaves with browner, softer shale. But as we're at the seaside, what we'd really like to find is a fossil.

Can this thing really be a seashell? It's made of stone, not shell. And it's not like the beautiful picture in the book, the one displayed at the museum, the one for sale in the shop. It's a smudged sort of thing, worn down by the sea. Maybe it's just some random sort of mark?

Aristotle, Greek philosopher from the third century BC, was aware of objects strangely like seashells found on the tops of mountains. The Roman poet Ovid puzzled over them in iambic pentameter: 'I have seen what once was solid earth change into sea, and lands created out of what once was ocean. Seashells lie far away from ocean's waves,

ABOVE Honeycomb weathering in Old Red Sandstone at Portishead, Bristol. This is caused by sea water evaporating. Salt crystals, as they form, expand within the rock and crumble it apart.

BELOW Layers of the Lias: alternating sandstone and shale at East Quantoxhead, Somerset. The blackness of the shale indicates a sea bottom deep enough to be deprived of oxygen.

RIGHT Two rock markings strangely resembling scallops, Hummersea near Saltburn, Yorkshire. Are they random freaks of nature; or somehow the same as the three living barnacles alongside?

and ancient anchors have been found on mountain tops.' (*Metamorphoses* Book XV). Shame he added the fanciful bit about the anchors.

For Aristotle, finding seashells in stones was no more unlikely than finding them in the sea. Oysters may be aphrodisiac but don't themselves enjoy orgasm, that's obvious. Shellfish don't have sex; such non-copulatory creatures arise by spontaneous generation out of suitable mud or slime, like flies from rotting meat. So why not some spontaneous generation inside a stone equally as under the sea?

But by the late Middle Ages, that wouldn't work. Fossils as spontaneously generated sea creatures didn't fit with the facts as recorded in the Bible. The seashells could not have been placed within the rocks on Day Five of the Creation (Thursday), as the earth and stones came into existence on Day Three (Tuesday).

Surely it was more reasonable to identify fossils as oddly-shaped stones, like the suggestive vegetables collected by Esther Rantzen on the TV programme *That's Life*? The ones you see in museums are specially selected. Fossils in the real world are unconvincing lumps and smudges. If you only knew carrots from Esther Rantzen, you might conclude that carrots were vegetable body parts, somehow seeded off some over-active gardener.

The fact is that stones resembling seashells *are* found on mountaintops. There are three possibilities. They are either:

- stones that just happen to look like seashells
- seashells left there from Noah's flood
- seashells, and we just don't know how they got there.

Leonardo dismissed Option Two. Leonardo knew how fast a clam can slither: 'It would not have travelled from the Adriatic Sea as far as Monferrato in Lombardy, a distance of 40 miles, in forty days and forty nights.' Further, he noticed that particular fossils occur at particular levels within the rocks. A single Biblical flood would surely have mixed them all together. On the idea that the molluscs had been generated within the rocks by plastic forces, he pointed out their growth rings. How could they grow, without feeding themselves; and how could they feed encased in solid rock?

Bravely, then, Leonardo embraced Option Three. 'Since things are far more ancient than letters, it is not surprising that in our days there exists no record of how the aforesaid seas extended over so many countries.'

Which brings us back to Nicholaus Steno, and his healing *Glossopetri* or tongue stones from Malta. They were found in the fields after heavy rainstorms: did the lightning knock them down out of the sky? A better explanation was in the Bible, in the *Acts of the Apostles* (Chapter 28). St Paul gets shipwrecked on Malta in AD 60, lights a fire, and a serpent crawls out of the firewood and bites his hand. Paul 'should have swollen, or fallen down dead suddenly', but was miraculously undamaged. Legend (but not the Book of *Acts*)

suggests that the saint then cursed all of the snakes on the island, causing them to lose their eyes and tongues. Tongue stones are shaped slightly like viper teeth and the Biblical evidence accounts for their power against poison.

According to Aristotle, an arrow flies in a straight line until it runs out of motion, then drops like a stone. And a stone behaves the same. Circles are perfect, so the world is necessarily a perfect circle, and the planets, sun and stars circle the world in perfect circles of their own. Aristotle wrote whatever he felt, then argued down anyone who disagreed.

We still use this method for religion and politics. But around the year 1650 there started to be a different way for science. Bold assertions of fact, even those of Aristotle, even conceivably those of the Bible itself, could be disproved by careful experiment.

So when Galileo suggested that the sun stayed still, and the earth went in circles around it, you believed it because Galileo was such a groovy guy. But more than that: any other mathematical genius could get out his abacus and Tycho Brahe's planet positions, and calculate Kepler's ellipses to see them fit the data. If the Italian chap turned out to be talking tosh, there were contributors to the Royal Society's *Philosophical Transactions* ready to point out the fact. (As well as a couple who'd point out what tosh it was – when Galileo had actually got it spot on.)

As Sherlock Holmes himself so aptly put it: 'The difficulty is to detach the framework of fact—of absolute undeniable fact—from the embellishments of theorists and reporters.' You can go for the philosopher with the fancy beard. Or you can argue it out in the *Philosophical Transactions* while someone in Leipzig tries to re-do the experiment. The second method has continued to the present day because of the surprising extra facts that it's turned up for us – plus the way that so many of those facts turn out to be actually true. Examples are the circulation of the blood; the moons of Jupiter; the microscopic living beings in a drop of water; radio waves, the transistor and the laser.

So what is a fossil?

If God hadn't wanted us to believe in evolution, he wouldn't have arranged for dead animals to be preserved in the rocks. And preserved, what's more, in a selection of different ways. What's a fossil? My dentist asked me this, and in the brief interval before he blocked my mouth with a blue rubber membrane, I was able to say that fossils are roughly the same stuff as the teeth he was about to drill into.

Fossils are found in sedimentary rocks, and most sediments occur under the sea, so most fossils are fossil seashells. Seashells are made of calcite, which is the same mineral as limestone; but it's limestone with a subtle stiffening of protein that modern materials scientists would be delighted to be able to use too.

Subjected to slightly acid seawater, seashells dissolve. But in rocks that aren't acid, seashells can stick around for 600 million years. Bones didn't exist 600 million years ago, they only go back to the Ordovician. But for the 500 million years they've been around, they also have been falling into sediment and getting buried before the bacteria could get at them.

That's in non-acidic water. If the water is acidic, that can dissolve away a shell while leaving the surrounding stone. The shell-shaped space refills with fine sediment, which then hardens. Two hundred million years later, a fall of the cliff splits it open onto a beach, and reveals an imitation seashell made of silt.

That gives you your fundamental fossil. In a more complicated process, a mineral like silica can seep into the buried seashell and replace, molecule by molecule, the original calcite. Silica is the very tough stuff that sand is made of. Silica fossils, supposing you should find one, are uniquely fine in their detail. One form of silica is opal: opalescent tree trunks, when you see one in a museum, are strikingly surprising but also strikingly lovely to look at. Another form of silica is flint: some flints contain beautiful sea urchins.

Another mineral, pyrites, can do the same trick. Pyrites is iron sulphide, whose crystal version is known as fool's gold. It creates a black fossil with a brassy sheen, sometimes contrasting nicely with the same fossil shape in pale brown, made out of limestone, above the same Yorkshire beach.

LEFT The typical fossil is only moderately convincing: ammonite fragment from Saltburn, Yorkshire

RIGHT Shelly sandstone beach pebble, Saltwick. This one is original shell, still preserving (not very clearly in the photo) its original mother-of-pearl sheen

If you bake bread in a traditional oven, you can't smell when it's ready. The least-done loaf goes back in for an extra ten minutes, to come out again, two days later, perfect in shape and texture, apart from being rock solid and black. You chuck it on the compost heap. Six months after that it comes back out of the compost, still looking exactly the same. Carbon is food for all living creatures, including the all-devouring bacteria. But pure carbon, as in that two-day-old loaf, is longer-lasting than any plastic debris. It can last for millions of years. Plant debris, washed into the sea and buried away from bacteria, can reduce to black carbon. That's another form of fossil.

Nature doesn't just want us to see seashells, bones, and black carbon. Sometimes she shows us the whole thing. Amber seeps out of pine trees and engulfs an entire mosquito. Ground sloths and giant elk come to drink at a pool that's actually, just under the surface, a tar pit. They fall in, and are found 10,000 years later in what's now suburban Los Angeles. Yoho, in the Canadian Rockies, should be renamed 'Oho!' (in fact that's roughly what 'Yoho' means in the original Cree Indian language). At Yoho at the very start of life on Earth, a process not understood converted organic carbon into clay minerals, preserving not just the shells and bones but the entire bodies of strange sea-bottom worms.

Fossils are found in sediments, most of which are from under the sea. The most easily fossilised material is calcite, that is to say, shell. Accordingly, most fossils are seashells. Evolution came up with the idea of calcite body-parts – shells – 550 million years ago. That point ends the Precambrian, the 4,000 million years when rocks were boringly fossil-free, and inaugurates the first of the eleven geological periods.

The molluscs – bivalves, snails, squids and ammonites – are just one phylum out of the 30 or so phyla making up the animal kingdom. The brachiopods, an almost-extinct group of shells, make a second and quite separate one (brachiopods are more closely related to us than they are to other seashells). A fossil that isn't a mollusc or a brachiopod is especially welcome, as an entry into the other 28 sections of life's magic cabinet. Organising life so as to make sense of it is a harder job even than organising an office filing cabinet. Actually, the life-grouping 'phylum' is from a Greek word for a tribe, while 'file' is from the French for a piece of wire for stringing documents together on.

Incidentally, it's mollusk in America, mollusc in the UK and the rest of the world. Both mollusc and mollusk were formerly used in the US. This suggests that mollusk is a later spelling, evolving spontaneously out of mollusc somewhere west of the Atlantic. That suggestion would be incorrect. Mollusk is the original, with mollusc evolving out of it in Britain, and the two together migrating to America – where mollusc found a language landscape hostile to words ending in 'c'.

You can trace species of mollusc through the Lias shales of the Jurassic – or spellings of 'mollusc' through the geology texts of the 1830s. Mollusc, mollusk and the French mollusque share a common ancestor in the scholarly Latin molluscum. But mollusque – according to the Oxford English Dictionary – is the common ancestor of both the other two.

Like Asian tree frogs reaching Australia on a floating log, a group of molluscues arrived in England in 1817 between the covers of Cuvier's great work *Le Règne Animal*.

Molluscs have a head at one end, a foot at the other, and everything else in between. There are three main sub-categories. Cephalopods ('head-feet' including octopuses and nautilus seashells) have left-right symmetrical bodies just like us, and sport a set of tentacles. Gastropods ('stomach-feet') include snails, slugs and limpets. Bivalves are double-shelled seafood with the two shells symmetrically matched: such as scallops, clams and mussels.

That quite separate phylum, the brachiopod ('arm-feet'), has top and bottom shells which are each symmetrical left-to-right, but which don't match one another. They probably

LEFT Cast of belemnite, Saltwick

CENTRE Belemnite reproduced as a solid calcite object, Bearraig Bay, Isle of Skye. The small star-shapes are bits of crinoid stem.

RIGHT Preservation as pure carbon: a nondescript plant fragment (3 cm across) from Crook Ness north of Scarborough

BELOW A trace fossil can be as exciting as a dinosaur footprint (Chapter 9) or as humdrum as a wormhole. Whatever creature made these wiggly burrows in Cretaceous limestone at Burnmouth, Berwickshire, crawled horizontally (parallel with the now upright bedding planes). So this was probably a sediment scavenger, rather than a filter feeder like a lugworm which burrowed up-and-down.

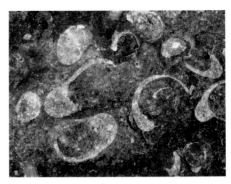

escaped from whatever wanted to eat them by clapping their shells together, to squirt water out backwards and propel themselves in the opposite direction. Brachiopods were the most successful seashells from the start right through to the mass extinction at the end of the Permian period. But that extinction hit them hard. Today a couple of hundred species survive, none of them on British beaches.

Shell is one of evolution's outstanding inventions: a simple sort of stone, reinforced with protein so as to be not just many times stronger than the original but also slightly flexible, and so far more resistant to shattering. The shelly sort of life hit the world at the start of the Cambrian period. That first creature exploiting calcite structures for body support and protection was so successful at eating others and not itself being eaten that its descendants diversified and spread into every salt-water lifestyle niche world-wide.

The insect equivalent, chitin, is a sort of sugar, stiffened with protein and (again) with calcite. Nature invented it soon afterwards: it's lighter than calcite, and its users – including today's crabs, lobsters and insects – are very happy with the stuff.

But the disadvantage of both of them is that they don't grow. Bone, invented by evolution in the second geological period (the Ordovician) 100 million years later, is better because you can carry it around inside you and it gets bigger as you grow up. (Well, we vertebrates would say that, wouldn't we? A billion insects beg to point out that bone is such subtle stuff that it takes a lot of our blood just to maintain it in working condition, and we can't wear it on the outside at all. Our only open-air bones are our teeth, and look what trouble they give us.)

Insects (and trilobites and crabs) get bigger by shedding their chitin shell and starting again. Seashells grow like houses, by adding bits on. Bivalves and brachiopods add new calcite all the way around the rim: each year, the same shape but slightly larger. More ingenious is to shape yourself as a spiral. Gastropods (snails and whelks) add new shell at the opening only, and get bigger while staying the same shape.

While brachiopods and bivalves are commonly (and wrongly) lumped together as 'seashells', the gastropods are recognisable by their curly spirals. But within the group of cephalopod molluscs, there are two more shells that are ex-citingly different. A belemnite is bullet-shaped. This is an

internal shell of a complex, squid-like animal, but as a fossil it's featureless enough to have been considered the result of a lightning strike, and named accordingly as a thunderstone. Well, they were sometimes found in the fields just after storms. (Readers could come up with an alternative theory of why belemnites and Maltese tongue stones should be so often found after heavy rain.)

Even more striking, though, are the spiral coils of the ammonites, lying below the cliffs of Cleveland and all along the Dorset shore.

Life and times of the ammonite

The original Ammonites descended from an incestuous re-lationship between Lot and his own daughters after the de-struction of Sodom and Gomorrah. Lot's wife, we are told in the Bible, suffered a theo-geological transformation into a pillar of salt in that debacle. Ammon was present-day Jordan, and the Ammonites were among Israel's many miscellaneous enemies. King Solomon, in the interests of ethnic harmony, married an Ammonite.

TOP LEFT Crinoids (known as 'sea lilies', but actually coral-like animals) in Carboniferous limestone, Barns Ness, east of Edinburgh

TOP MIDDLE Dinosaur backbone, Whitby Museum

TOP RIGHT Shelly limestone, Robin Hood's Bay, Yorkshire

LEFT Carboniferous colonial coral, stone 20 cm across: Borron Point, Dumfriesshire

BELOW Carboniferous coral, and a few shells

RIGHT Magnificent ammonite selection, the largest about 50 cm across, in Whitby Museum

BELOW Large ammonite and Lias cliffs, Traeth Mawr, Glamorgan

Pliny the Elder (who died in AD 79 while observing the eruption of Pompeii) named the curly fossils ammonis cornua, or horns of Ammon, because the Egyptian god Ammon wore ram's horns. Ammon was an important creator-god, and the emperor Tutankhamun was also named after him.

In the Middle Ages they were 'serpent stones', the petrified remains of the snakes St Hilda drove out of Whitby to make a safer sort of seaside resort. Since the specimens found never seemed to have their heads still on, the finders often carved replacement snake-heads.

The first ammonoids – the shells called bacrites from 415 million years ago – were straight, not spiral. Ammonoids flourished during the Permian, but at the start of the Triassic only one species survived the extinction called the Great Dying. That species ramified to fill the oceans of the Jurassic. Its descendants are the true ammonites; they took over the lifestyle and looks of the preceding ammonoids. (In the same way, after the extinction of all ammonites, the present day nautiloids have taken on their looks and lifestyles.)

Ammonites are sometimes found in worm-free, infertile sorts of sea-bottom strata. This implies that the ammonites have just dropped in, and were free swimmers in the ocean overhead. They lived in their outermost chamber, and used disused former homes back down the shell as flotation devices, pumping water in and out through a small tube – modern submarines use the same system for going up and down. Going by their modern relatives, ammonites proceeded through the sea upright and edge-first, with just an eye and a set of tentacles emerging from the shell end. The efficiency of their swimming shape has been tested in hydrodynamic tanks. It's a tribute to the overall usefulness of calcite shell that even these jaguars of the Jurassic were prepared to haul around such big chunks of the stuff.

Ammonites squirted ink at their enemies in the same way as their relatives of today, the squid and octopus. They had many tentacles, which they possibly used to seize their prey before swallowing them whole. More certainly, ammonites made a tasty snack for any passing plesiosaur. The plesiosaur in Whitby museum had enjoyed a final snack of four or five ammonites.

Jurassic ammonites turned out to be quite bad at evolving. Every million years, the current ammonite tended to become extinct and a new one moved in. This, combined with their free-swimming lifestyle, makes ammonites extraordinarily useful markers of time within the rocks. Of the 74 ammonite bands identified within the Jurassic, 71 can be found at Lyme Regis.

Towards the end of the Jurassic, the seas started to bring forth monster ammonites the size of car tyres. Ones of 50 cm are found in the upper Jurassic at Portland and in the Cretaceous at Peacehaven near Brighton. Ones 2 m across

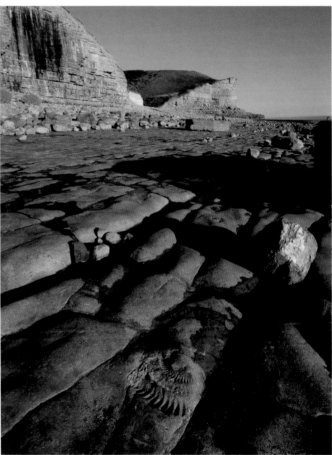

occur in Germany, and also in Whitby museum. Why so big? It might be possible to trace a connection with the hugeness of dinosaurs at the same time, and with the high carbon dioxide level and global warming of the Cretaceous, which must have been helpful to plant life.

A more conventional explanation points out that an ammonite couldn't eat anything larger than her shell-end opening. Some crab, shrimp or sea urchin was available to eat, and no dog-size plesiosaur was around to take advantage. Why did the ultra-ammonites became extinct again? Presumably something else arrived, able to eat these big dinners, and not burdened by tens of kilograms of calcite shell.

Ammonites died out in the extinction at the end of the Cretaceous which also did in the dinosaurs. If, as currently believed, that extinction was caused by a meteor strike; and if, as also believed, the ammonite had a free-swimming plankton stage in the sunlit surface waters before developing its shell; then the years of darkness following the meteor can be blamed for doing in all the ammonites.

The sediments of Tuscany are fairly recent. To Nicholaus Steno their fossil clams, oysters and bivalves were obviously clams, oysters and bivalves. In England, on the other hand, the commonest fossil is an ammonite from the Jurassic. An ammonite isn't any of the seashells of today. So when Steno and other Italians wrote that fossil seashells look just like seashells, the English just muttered 'silly foreigners'.

Sandstone and sainthood

So what of Steno's follow-up, the 'Postdromo', the actual dissertation on the strata? The sad fact is, he (once again) got distracted. After the strict simplicity of Lutheran Denmark, the supermarket shelf of conflicting religions in Holland went to his head, and indeed his soul. Steno became a Roman Catholic, and more than that, a bishop – he would have preferred a life of obscure self-deprivation as an ascetic priest. The Church sent him to the small group of Catholics in Hanover, Norway and his native Denmark – sand-covered, rockless Denmark. Disheartened by the bureaucracy and corruption of the Church, he turned to fasting and self-denial, and died in 1686, aged 48.

But Steno's story doesn't end there. Lutheran Denmark suffers a shortage of Catholic role models. In 1938, Danish Catholics petitioned Pope Pius XI to start admitting Steno as a saint. In 1953, Steno's remains were transferred to a splendid coffin and a special chapel in Florence's San Lorenzo. In 1988, John Paul II confirmed Steno as one of the 'beatified', a sort of probationary saint, to be known as the Blessed Nicholai Steno. Under current regulations, full sainthood requires two confirmed miracles. Steno has achieved the post-mortem miracle – a cancer cure achieved after prayer to him. But the lifetime miracle may be hard to uncover, for the man who wrote (after discovering that the heart was a pump not an oven): 'If they can be so mistaken about material things … which in an hour or so I can get a ten-year-old to demonstrate … what certainty do they give me that they are not also in error when they state their views about God and the Soul?'

Whatever we think of Steno as the only geologist even half-way to sainthood, we have to admire his clear-sightedness. It's not easy to see the obvious when nobody has ever seen it before and even Aristotle knows it isn't so.

When we look at the stones we call sedimentary, we find them to be made of other, earlier stones – with the odd seashell and bit of tree fern. Sandstone is made of sand … and sand is made of sandstone. As James Hutton put it in 1795: 'We see no vestige of a beginning, no prospect of an end.' This could lead to a theory of rock origin that is dead simple, but unsatisfying. The world is just made of bits of itself, recycled. Rocks don't have any origin: they are made of previous rocks. Which themselves are made of – previous rocks. Sandstone is made of sand: sand is made of sandstone. And so on back, for ever and ever.

Thank goodness for the granite. It means we don't have to adopt a theory so philosophically minimal. The group of rocks called igneous is not composed of other earlier rocks. Granite and gabbro, basalt and dolerite are, closely examined, crystalline. How they got there was the red-hot topic of the 1780s. Literally, in the case of James Hutton, who reckoned that the granite had arrived in a molten mass from underneath. Everyone else, however, knew that crystals start off as stuff dissolved in water, which made sense as the granite and the basalt had clearly arrived by the evaporation of Noah's Flood.

And just to prove it, at Portrush in Northern Ireland there was basalt with ammonites in it.

TOP Ammonite in limestone, showing internal chambers

BOTTOM Plesiosaur (marine dinosaur) in Whitby Museum – and its last ammonite snack

Stones with stripes

The world is made of three different sorts of rocks, which come into existence in three quite different ways. Most of the UK coast happens to consist of **sediments**: rocks made out of the broken down dirt and pebbles of other, earlier rocks. There are the squashed rocks, the **metamorphics**, also made out of earlier rocks, but by heat and crushing. Then there are the **igneous** rocks made from fresh magma, cooled down and crystalline.

Bedding

Sediments are where we start. Sediments are easy to recognise because they're laid down in beds. Whether under the sea, under lakes and rivers, or formed of desert sand, they come in regular layers. From afar they are seen with parallel stripes on.

Sadly, some sedimentary stripes are something else. Get this wrong, and you end up with rock nonsense. This section summarises the different sorts of stripiness in stone. More detail will be in later chapters.

Metamorphic rocks: cleavage

These rocks are not sediments and these layers are not sedimentary beds. The rocks are slates, formed by compression deep underground. In other words, they are metamorphic. The layering is called 'cleavage', and forms at right angles to the direction of compression. Whatever bedding the rock had to start with is usually, but not always, obliterated.

ABOVE Sandstone, Whitesands Bay, Pembrokeshire. The stripy colours follow the bedding planes. Otherwise, the steep angle and the smoothness of the layering might lead you to mistake this for slates.

LEFT Barlocco Bay, Dumfriesshire. Bedding planes divide strata that are alternately soft, and softer.

RIGHT Beds of grey shale and brown limestone alternate at East Quantoxhead, Somerset. Sediments, as the water is squeezed out by overlying rocks, can shrink; and the limestone layer has developed vertical joints in the same way as the granite and basalt on the following pages. One clue, though, makes this one certainly a sediment – the fossil ammonite fragment.

ABOVE Slate sea-stack, Abereiddy, Pembrokeshire

BELOW This slate cliff shows both bedding, roughly horizontal, and slaty cleavage, nearly vertical. Hole Beach, Delabole, north Cornwall

BOTTOM Chalk with parallel fault planes, White Rocks, Antrim

There are some clues to cleavage. Slate's cleavage planes are flat and smooth. They are spaced throughout the rock: you can split the stone anywhere. If there's any colour banding or texture change, it doesn't follow the lines of splitting, but lies across them.

You can wander right through this book to its final chapter without meeting any metamorphic rocks. In that unlucky 13th chapter on ancient Pembrokeshire, dark, solid sedimentary sandstones are (despite the clues just listed) confused with dark, solid metamorphic slates.

Faulting
Occasionally, parallel fault planes can mimic bedding or cleavage. Unlike cleavage, faultlines don't obliterate existing bedding. Most faults show shattered rocks along the join.

Igneous rocks: columnar jointing in basalt
Lava-flows are too lumpy to be mistaken for sedimentary beds. But when hot rocks cool down, they shrink. The shrinkage cracks are called joints. (They are different from faults as the rocks either side of a fault have moved past each other.) Columnar jointing, seen most famously at the Giant's Causeway, can look like upright bedding, or like slaty cleavage. Seen from closer, it's formed of columns rather than parallel planes.

Igneous rocks: horizontal jointing in granite
And the next set of layers? Opposite is granite at Land's End. Granite is an igneous rock. It starts off as a great molten blob deep underground, and solidifies while it's still down there. As erosion removes other stuff from on top of it, granite expands. As it expands, it cracks apart, as an onion would if you suddenly placed it in outer space.

So these are another sort of joint, and in uniform granite they develop in all three directions. The horizontal jointing could be mistaken for bedding. It isn't, because you already identified the granite from its speckled crystal structure.

ABOVE Columnar basalt, Dunbar harbour

LEFT Columnar joints, but not very clear ones, under Lindisfarne Castle. In the foreground, column tops. Both are parts of the Great Whin Sill

BELOW Land's End granite, Pordenack Point

3. ANTRIM BASALT

Giant's Nose, and Dunluce Castle on its basalt vent. In the distance, Benbane Head rises above the Giant's Causeway shoreline.

3. ANTRIM BASALT

The dotted urchin's studs the cliffs adorn
And blue basalt is stamped with Ammon's horn [i.e. with ammonites].

William Drummond, in *The Giant's Causeway* (1811)

All of England, apart from Cornwall and the Lake District, lies over sedimentary rocks. The sedimentary rocks, the sandstones and shales, came from the crumbling away of whatever was there before.

Elsewhere in Europe, in the eighteenth century thoughtful men were beginning to be dissatisfied with the idea that rocks just came from even earlier rocks. It all had to start somewhere. And where it started had to be the granite and the basalt, rocks that weren't patted together out of leftovers, but were crystalline in nature. Granite is clearly crystals; basalt is crystals too, when looked at closely through a lens.

Crystals come when stuff is dissolved in water, and then the water boils away. That's how they got the alum crystals shown in Chapter 9. You can do it with sugar, or salt, or copper sulphate. In the same way, granite and basalt crystallised out of a universal ocean that once covered the whole world. This made sense, as such oceans had indeed occurred on at least two occasions. Once, with Noah floating on top of it: '…the waters prevailed exceedingly upon the earth; and all the high hills, that were under the whole heaven, were covered.' (*Genesis* 7. 19). And before that, right at the very start, when 'the earth was without form, and void; and darkness was upon the face of the deep. And the Spirit of God moved upon the face of the waters.' (*Genesis* 1. 2).

This theory of the origin of the original rocks was called Neptunism (after Neptune, god of the ocean). It was formulated around 1775, by the talented German mining academic Abraham Gottlob Werner.

BELOW Columnar basalt at Benbane Head, above the Giant's Causeway

TOP RIGHT Portrush from the west, with the Portrush sill displaying its columnar joints. The argumentative ammonites are on the other side of the sill.

RIGHT Ammonites (about 5 cm across) near Coastal Zone countryside centre, Portrush

BOTTOM RIGHT The ammonite site at Coastal Zone, Portrush

The various rocks could be divided, bottom to top, into three fundamental layers. Primary rocks, like granite and basalt, crystallised out of the first ocean. Secondary rocks eroded out as the sea sank and the first mountaintop islands emerged; those secondary sediments sagged in folds and ridges across the still half-submerged mountains. Finally, the Tertiary rocks are soft, level-lying sediments laid in flat layers over the lowlands during the final retreat of the seas. And as Noah's Flood and other such disasters were world-wide, within these three main layers are various 'Universal Formations' – the chalk, the Old Red Sandstone – laid down at the same time all over the world.

But then in 1788, James Hutton and his young friend John Playfair visited the Berwickshire coast, as described in the Introduction to this book. There they saw that sedimentary rocks had arrived at least twice, and with a huge time gap in between. Hutton was convinced that sedimentary rocks had formed at the sea bottom not twice, but again and again and again. And when he examined the granite, that first and most fundamental sort of stone, he found it apparently arriving in a molten state, squeezed into the gaps between existing sediments. It was at Glen Tilt in the Scottish Highlands that he came across the granite so squeezed, and on Arthur's Seat, in the middle of Edinburgh, the same thing had happened with basalt. The first rocks came not from water, but from fire somewhere underground. Hutton's theory was called 'Plutonism', after the ancient god of the Underworld.

Even earlier, in 1771, Nicolas Demarest had used his job as Inspector-General of Manufactures in France as an excuse for geological excursions. He found basalt in the Auvergne, lying where it had flowed out of an obvious volcano (the Puy de Dome, now known to be less than 10,000 years old). And it looked the same as the just-published tourist engravings of the Giant's Causeway.

But what could a Frenchman know about Irish rocks? In 1786 an Irish cleric, the Revd William Richardson, put paid to the ungodly Plutonic hypothesis. He discovered the fossils of ammonites, sea creatures, within the Antrim basalt. If it's got fossil seashells in it, then basalt comes from the bottom of the sea: this is beyond all dispute.

Well, geologists are never beyond all dispute, where would be the fun in that? Hutton's Unconformity in Berwickshire is pretty believable, and so are the mountains of the Auvergne. But then, so are the ammonites in the black Antrim rocks, when you come across them just north of the Coastal Zone countryside centre in Portrush. Study the fossils there, and the rocks they're found in, spend a month or two exploring the Antrim basalt and the chalk it lies on and under, and work out what exactly is happening here. But supposing you haven't the time for that, Hutton's young friend John Playfair came over and did it for you in 1802.

The first thing he realised about the black basalt of Portrush was that it was different from the Auvergne and the Giant's Causeway, and it wasn't a flow of lava. Instead it was a sill: a layer of red-hot rock squeezed between two pre-existing rock layers. The smooth, flat top of it shows that it was inserted into the space between two beds underground, rather than having been a lumpy red-hot river out in the open air.

The basalt is Antrim's most recent rock, barring the clay and boulders left behind by the glaciers. That ought to put it on top of everything else; and on top of everything else is where, on the whole, it finds itself. But if the Portrush basalt squeezed in underground, this breaks the rule. What's on top

of the basalt is what it squeezed in under, and that is necessarily older, and could be quite a lot older. And looking again at the topmost black rocks, they do seem unlike the rest. There are what look like bedding planes, as if it were sedimentary.

That rock on top certainly isn't any Antrim chalk, as it's quite the wrong colour. But underneath the chalk is a layer called the Lias, brown sandstone and mudstone and shale. The Lias is well known in Dorset and Yorkshire, but is only spotted above sea level by the sharp-eyed at a couple of places along the Causeway Coast. This top rock at Portrush that's sedimentary-layered, but black and basalt-hard, doesn't look like any Lias sand and shale. But remember, the basalt sill came in underneath it still red-hot. This black, brittle, but layered-looking rock is Lias shale that's been baked by that arrival. Like a cake forgotten in the oven, it's black and rigid. That cake still carries the inedible outline of its currants and glacé cherries; and in the hornfelsed Lias, you still can see the little black ammonites.

Even so, nine years after Playfair's visit, William Drummond was writing his rock lyric quoted above – it rhymes, so it *must* be true! Even if he does have an odd way of pronouncing 'basalt'.

Giant's Causeway

Boswell: 'Is not the Giant's Causeway worth seeing?'
Johnson: 'Worth seeing? Yes, but not worth going to see.'

James Boswell, *Life of Samuel Johnson* (1791)

FAR LEFT Looking north from the Giant's Causeway. Chimney Tops, a bundle of isolated basalt columns, on the skyline. The red band of rotten rock called laterite divides the Lower Basalt from the overlying Causeway Basalt.

LEFT So-called Giant's Eye in the laterite layer. Oxygen-rich groundwater seeped into the basalt along cracks and joints, leaving some lumps of rock in the middle relatively less damaged.

BELOW LEFT Lava-flows near Rinagree Point, Portstewart. The base of the upper flow has been shattered by steam from underneath, when the red-hot lava covered either wet ground or shallow water.

BELOW RIGHT Having studied the magnificent Giant's Causeway basalt, it's worth taking a glance at the rest of the UK. Underneath all the nesting guillemots, a reddened laterite layer marks the top of one lava-flow and the foot of the next at St Abbs Head, Berwickshire.

Nearly a million people a year disagree with Dr Johnson, including me. But even if you are an irritable dictionary-compiler in a scruffy wig, there's more to the Giant's Causeway Coast than just the Giant's Causeway. However, the Causeway is the place to start. Of the UK's 25 UNESCO World Heritage sites, just two are 'natural' rather than cultural – the Jurassic coast of Devon and Cornwall, and the Giant's Causeway. The Giant's Causeway is Britain's most visited bit of geology, apart perhaps from Cheddar Gorge. Cheddar is cheesy in both senses of the word, but the Giant's Causeway is simply over-the-top. Basalt lava-flows typically have red-stained bits between them. At the Giant's Causeway, that red-

staining is 4 metres deep. Basalt lava-flows have columnar jointing, a vertical breakage pattern caused by cooling. The Giant's Causeway has 40,000 such columns, so perfectly polygonal they're clearly the work of a Fingalian superhero.

Basalt lava happens in all sorts of places. Wherever you see it, it's like the Giant's Causeway, just not nearly so much.

Basalt is tough stuff when it's underground. But up in the open air, it gets attacked by dangerous alien chemicals, such as air, water, and rotting plant juices. If you lived inside an active compost heap for a million years, it might damage your complexion too.

The first basalt flowed out over Ireland and lay under the

rotting plant life of 50 million years ago, when the UK's bit of continent was a lot further south than it is today. The air, water, and plant juices attacked it, and turned the iron inside it bright red. A red layer formed the top of the first basalt, and then the second basalt flowed over.

For a long time, nothing much happened. A river flowed over the lower basalt, and carved a valley in it. And over long, boring millennia, the top of the upper basalt was broken down into a red soil and crumbly gravel, many metres deep. Fifty million years later on, this red rotten rock was an easy way to carve out the tourist path, which is why it now shows up so satisfyingly on pictures taken from the west.

But then it all started up again. A nearby lava vent blocked off the river valley, and then spread a lava pool right across it. Basalt lava is, as lava goes, fluid and slippery – it's more like treacle than tar. It formed a smooth, level pool right across the valley. It solidified, and started to cool.

Rock that cools must shrink. Shrinking in the downwards direction is easy enough, but shrinking inwards isn't. Rocks are stiff, and when they cool they crack apart. When the cooling rock forms a smooth regular layer, it cracks apart in a regular way. Specifically, it cools in hexagons, like a honeycomb.

The link between bees and basalt is a matter of geometry. Human beings, faced with the problem of breaking chocolate into bite-size pieces, separate it into squares. Squares and right-angles suit our machinery and our simple way of looking at things. But Nature is more efficient. Given the problem of dividing a hive space into cells for your grubs using the least amount of wax, bees come up with the correct answer, which is hexagons.

The basalt has the same question, in reverse. The energy to pull apart a piece of rock depends on the area of the broken surface. The more new broken surface you create, the more work you do. Nature is lazy, so given the problem of breaking up a slab of basalt, it does so by creating the least possible amount of broken surface. Again, the hexagons are the correct solution.

Bees are better at it than basalt. The hexagons of the beehive are almost perfect. But as the basalt cools, it's as if tiles were being laid by a dozen different workmen starting in different parts of the bathroom. You end up with quite a lot of five-sided and seven-sided shapes.

They say that knowing the science ruins the romance. The romantic version of the Giant's Causeway is that Finn McCool decided to have a fight with a Scottish giant called Benandonner (who, from his name, may actually have been

FAR LEFT Even after the columns had formed, differences in temperature within each column meant it had to crack apart horizontally as well. The exact mechanism of the 'chisel marks' seen here isn't understood.

LEFT The 'other end' of the causeway: columnar basalt on the island of Mull. The columns were bent by being nudged sideways before fully cooled. A similar structure, the 'Giant's Harp', is one bay east of the Giant's Causeway.

ABOVE Magnificent basalt at Waterstein Head, Isle of Skye. A reddened laterite layer can be made out about two thirds of the way up.

a mountain). But being over the maximum weight for the Larne ferryboat, Finn had to build the Causeway to get there. When he did get there, he wished he hadn't, because whether or not he was a mountain, Ben was a bit too big. Finn ran back across the causeway, wrapped himself in his giant blanket, and jumped into bed.

Benandonner came after, shouting: 'Where's this mighty Finn?' But Finn's wife, Mrs McCool, points to Finn in his bed and says: 'Shush, shush, we hate it when you wake the baby.' Well, if that was Finn's big baby, Benandonner decided he didn't want to meet its dad. He headed back to Scotland, pulling up the causeway behind him.

That's a pretty silly story. If Finn McCool had really been the builder, he would surely have done it the straightforward way, in squares.

Finn lived before the discovery of plate tectonics and the history of moving continents. Even so, as he ranged across Ireland and Scotland with his thousand Irish wolfhounds lead by white-breasted Bran, he gained a hillman's knowledge of rocks and scenery. In a dim and mythic way he was aware that his Causeway between Antrim and the Hebrides was simply rebuilding a geological connection of 60 million years ago.

Are the columns of the Giant's Causeway the very same lava flow as those of Fingal's Cave on Staffa, Scotland? It's possible: the chemistry and dating fit, at least approximately; and the Columbia River basalt in the northwestern US flowed in just a few days about 200 miles out of Idaho right across the state of Oregon almost to the present-day shoreline. The Columbia River has the same hexagonal columns as the Giant's Causeway, though not quite so photogenically displayed.

More importantly, the strange scenery of Antrim, with its low black cliffs, the flat-topped landscape with its fertile green grass, is geologically the same as northern Skye and most of Mull. All of it was formed a mere 60 million years

ago, when the rest of the UK landscape was already tired and old, with the opening of our most recent ocean, the Johnny-come-lately Atlantic.

The Atlantic today is growing at about an inch per year. Think back 50 million years, to when it was no wider than the Mediterranean – or earlier, to when it was the width of the Red Sea. And back before that to the moment of the big split, when the Atlantic Ocean cracks open as a little fissure across a black landscape. The fissure winks red in its grey depths, gives off steam and black ash and the occasional firework fountain of yellow sparks. Over a century or two it enlarges into a permanent rift stinking of hydrogen sulphide (the same gas as farts). A similar rift currently runs past the ancient Viking parliament or Thingvellir in southwest Iceland. Not just similar, it's actually the same mid-Atlantic rift, at the one place where it appears above sea level.

The combined Antrim-Hebrides-land, 50 million years ago, would have been like Iceland today, but with a nicer climate (in between the episodes of poisonous grey haze and acid rain). Low, rounded mountains of rubble and bare rock, all coloured in the ever popular plain black; rock pinnacles in round spindly shapes; scree slopes in mineral colours of greyish orange or pink; the occasional green pool with its stink of bad eggs. But in the hollows below the bare ridges, the reddish soil is rich with Mediterranean wild flowers. At night, red fountains of lava light up the horizon. And out to sea, beyond the black sand beach, a new black rubble island rises in steam and black ash.

If Northern Ireland and Staffa are one country geologically, they were so also socially in the clan times up until AD 1608. The sea between united, rather than divided, Argyll and Antrim. When spring stirred and you fancied a bit of intermarriage or maybe some neighbourly raiding, rather than trekking over the mountains it was easier to head south across the sea, in the man-powered galleys not much changed since they were Viking longboats.

There was one difference the MacDonalds of Skye and Argyll noticed when they came south. Antrim's north coast lacked the great ice-carved sea lochs of Scotland. The glaciers of Antrim had no high mountain hinterland. They were, as glaciers go, fairly puny, and didn't flow out northwards against the more powerful offshore ice-flow heading south. Antrim's glaciated bays (none fine enough to call a fjord) are round the corner at Cushendall and Carnlough, where mountains behind supply ice, and glaciers could edge out sideways into the southbound Irish Sea ice-flow, like cyclists at a T-junction. However, the north coast has one geological compensation. The stretching of the land at the time of the opening of the Atlantic has created faultline harbours like the one at Ballintoy.

LEFT Sea stack two bays west of Giant's Causeway. The right-hand half of the stack is formed from a dolerite dyke. Dykes cool sideways rather than up-and-down, so the columnar jointing in them is horizontal.

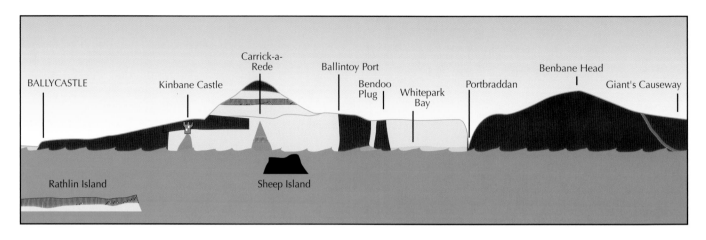

Antrim Coast: key

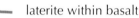 basalt: dolerite sill

 laterite within basalt

 chalk: volcanic vent in chalk

Jurassic: Lias

Heading south from their sea lochs, the MacDonnells (now spelling their name in Irish style) intermarried so skilfully as to inherit much of Antrim. Sorley Boy MacDonnell achieved financial security by kidnapping and then ransoming the constable of Carrickfergus Castle. From about 1550, he expanded his land holdings by defeating the MacQuillens. Unfortunately, this made him so important that his future battles would be with the Queen of England – he was attacked on Rathlin Island by Sir Francis Drake himself.

Like the Scottish Highlanders, the Irish tribes proved particularly resistant to rule from London. Twenty years after Sorley Boy's death, all land owned by the O'Neills and O'Donnells was confiscated by the British government, and repopulated with Protestant, English-speaking 'tenants', mainly from the Lowlands of Scotland. Antrim was similarly colonised, but in a free-enterprise fashion by wealthy aristocratic individuals rather than the state.

The rather predictable result was four centuries of repression, exploitation, misery and rebellion – which may finally, since the Good Friday Agreement of 1998, be coming to its end. On the day that I took these photos of the Giant's Causeway, the Savile Report on the Bloody Sunday Massacre finally accepted that those killed by British soldiers 30 years before had been unarmed, innocent protesters. And in a minor but happy effect of the Peace Process, returning tourists find along the Antrim coastline one of the UK's most charming and only moderately overdeveloped bits of seaside. Clifftop paths and beaches, sea stacks and arches, little seaside towns and enough ruined castles to satisfy even Sorley Boy MacDonnell.

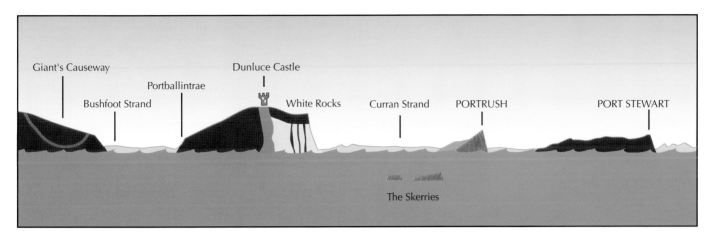

Giant's Causeway · Bushfoot Strand · Portballintrae · Dunluce Castle · White Rocks · Curran Strand · PORTRUSH · PORT STEWART · The Skerries

Vents Unfold

All along the Antrim coast line, basalt lies on top of chalk. But that volcanic basalt had to come up through the chalk somewhere. Apart from the Giant's Causeway, all the main tourist attractions and ruined castles happen to be on the various sorts of volcanic vents listed below. With black basalt coming up through nice white chalk, the colour contrast couldn't have been more helpful. The National Trust car parks are pretty convenient as well.

1 The Bendoo Plug, just east of Ballintoy Harbour, is a volcanic vent filled with tough black dolerite. Dolerite is basalt that's cooled underground rather than on the surface, and so is slightly crystalline. The vent is about 300 m wide, and rises from the sand flanked on either side by chalk. The actual dolerite/chalk contact can't be seen; the chalk that was right up against the cooling basalt has been cooked and so has crumbled away. But the chalk a few metres away appears unaltered by the intrusion, so geologists believe the basalt was flowing up this vent for quite a short time. It was possibly the supply pipe for a little lava lake.

Basalt is the less-lethal lava. In Hawaii today, Bendoo would have a viewing terrace, with a bar so as to let the light of the red-hot gently swelling lava twinkle in the depths of a rum daiquiri.

2 Dunluce Castle. The McQuillens chose badly when they built on this lump of vent agglomerate, at least according to the Girl Guides of Cullybackey (their useful and attractive leaflet is called 'Portrush Rocks!'). The vent agglomerate consists of basalt lumps loosely welded with volcanic ash – and in 1639 the castle's kitchen quarters crumbled into the sea, along with all seven of the cooks. They'd have been better off, the Girl Guides suggest, building on the basalt headland immediately to the east. Well, the McQuillens might have found that headland with its inland access more awkward to defend against Sorley Boy McDonnell in 1565. But they'd at least have had a nice hot supper after the battle.

ABOVE Vent agglomerate, Dunluce Castle

LEFT Bendoo Plug – the panoramic photo makes the curve of the vent more apparent than it is in real life.

After defeating the McQuillens anyway, Sorley Boy married his brother Colla to one of their eligible maidens, thus preventing some future battles. In 1584, though, the English attacked while Sorley Boy was away, and battered the castle with cannons for three days before it surrendered.

Sorley Boy tried to negotiate with the English, but when they proved unfriendly, he simply retook the castle. Queen Elizabeth decided to confirm this arrangement, and he remained as a reasonably peaceful Constable of Dunluce until he died.

3 Carrick-a-Rede's tidal island is unusual in not having a ruined castle on it. Instead it was the site of a salmon fishery, connected to the mainland by a rope bridge. The National Trust's ropework is considerably safer than that used by the original salmon netters. That said, just as you're going over, there's a young fellow behind you who likes to bounce the bridge.

Each end of the bridge is attached to a vent agglomerate which contains a mix of basalt lava with quite a bit of chalk. This gives it its greyish-green colour, and a scooped surface where chalky bits have dissolved out. The east part of

the island, and the mainland to the east of the rope bridge, are a dolerite intrusion, squeezed out sideways by the same volcano. A quarter-mile away in each direction, the chalk coastline resumes.

4 Kinbane Castle is reached down a mile of single track road below arching hedges. At its end, a council worker in his van eats a lonely lunch overlooking the ocean. A walkway leads down between brambles, around a steep corner and then by crumbling concrete steps. The castle below occupies an almost-island of shattered chalk, its seaward end hanging

Dunluce Castle, and the Girl Guides' preferred site on the basalt headland

impressively over the Atlantic. It's reached across a tideline of black and white cobbles, with today a handy iron ladder.

One of Antrim's less-visited visitor attractions, it was more of a place in 1555, when it was held by Sorley Boy MacDonnell's younger brother Colla against an English attack using cannon. During another attack Colla lit beacons on the seaward end of the castle, with its splendid sight-lines along the coast. His neighbours responded helpfully.

BELOW Carrick-a-Rede Rope Bridge: chalky vent agglomerate, and (beyond the visible end of the path) dolerite sill

BOTTOM RIGHT Chalky agglomerate, its chalky component partly dissolved away by rainwater, Carrick-a-Rede

The English were trapped in Lag na Sassenach, the Englishmen's Hollow, between the castle and the mainland, and successfully slaughtered.

The vent here was all gas and very little basalt. So the vent agglomerate is mostly lumps of chalk. It isn't obviously a vent at all – chalk could be shattered much the same by faultlines. But when you see lumps of basalt enclosed within the chalk – how else could they have got there?

Thanks to National Trust ownership, nearly 30 miles of coast path connect the Giant's Causeway, the four volcanic vents, five or six wide sandy beaches, and the coastal towns of Portstewart, Portballintrae and Ballycastle. The Causeway itself is best at evening, when the café closes and the buses

cease to run; then walk on, along 5 km of high cliff, looking down on the bay with the Spanish Armada shipwreck, and the bay with the sea stacks, and all the basalt columns. It helps not to know too much about sea erosion. The black cliffs are pretty well vertical; the sea's carving into the base; some day it's going to come down and somebody's going to be walking on top of it when it does …

The path drops to a contorted shoreline. At the back of the bay, rock pillars rise out of green grass, and there's an inland cliff with sea caves: Antrim has well-displayed raised beaches. Atlantic surf splashes like milk across black boulders. An eider duck chivvies a bunch of fluffy featherball babies, somehow adapted to survive between the boulders and the waves. Offshore are the mighty ruins of sea stacks and arches: some in black, some chalk white. The bay back cliffs are black and white in ways we ought to understand. But as well as the four volcanic vents, there are dolerite sills creeping in among the older rocks. And there are faultlines. Small Ballintoy Harbour has white rocks to port and black basalt to starboard: a later chapter will return to Antrim to explore this easy-to-see fault.

The coast path runs down to Ballycastle. Here Sorley Boy watched across the water as his family and followers died under the swords of Sir Francis Drake. But above the sea haze in the east rises the 200 m-thick dolerite sill of Fair Head. Out to sea, Rathlin Island is a two-layer affair, chalk lying under a six-mile slab of basalt. Antrim's history is tragic and complicated. But Antrim's geography is always black and white.

TOP Kinbane Castle. Two or three patches of black show where the chalk has engulfed some basalt.

ABOVE Mixed chalk and basalt in vent agglomerate, Kinbane Castle

4. HOT ROCKS

Heads of Ayr: the vent of a Carboniferous volcano. Lava-flows, ash falls, dykes and sills add igneous incidents to the UK's mostly sedimentary coast – 1 km south of this basalt headland are the handsome sandstones shown on page 166.

4. HOT ROCKS

The Andalusian merchant, that returns
laden with Cochineal and China dishes,
reports in Spain how strangely Fogo burns,
amidst an ocean full of flying fishes!
These things seem wond'rous,
yet more wond'rous I,
whose heart with fear doth freeze,
with love doth fry.

Thomas Weelkes (madrigal, *c* 1600)

During the morning of 24 August, a strange cloud appeared above the mountain. By 6pm the cloud was obviously coming *out of* the mountain: a column of grey ash that divided like the branches of a Mediterranean umbrella pine. Before nightfall, ash started to fall onto the town. The stuff is actually not ashes at all, but volcanic lava in flakes small enough to drift down like dandruff. By midnight, it lay several metres deep. Roofs started to collapse under the weight.

Along with the ash fell potato-sized lumps of pumice. Because of being mostly not stone but gas bubble holes, these were not too painful. Fleeing townspeople found that pillows over their heads would protect them, but a naval rescue force was unable to land because of the floating pumice that blocked the harbour.

The darkness was complete – 'like a sealed room in which the lamp has been blown out'. So they couldn't see the column of gas, almost red-hot, a kilometre wide, and rising to a height of 33 km, well into the stratosphere. Like a vertical flame-thrower, the gas shot upwards at hundreds of miles per hour, carrying ash flakes, pumice, and lumps of solid rock. The unhappy townspeople couldn't see it – but they could hear it: a continuous roar like an express train at rooftop level. Lightning flashed in the ash-cloud, and the earth shook.

At about 1am, those still in the town – and still alive – noticed the fall of pumice and ash ease off, and finally stop. This was, in fact, the worst news of all. The convection column above the volcano had become unstable, and collapsed. Its ash and pumice returned abruptly to ground level: and then – like a hovercraft on a gust of hot gas and groundwater steam – flowed down the mountainside at the speed of a fast car. About 20 minutes later, the airborne hot avalanche hit the town. Buildings were broken off at the level where they emerged from the lying ash. Those still in the town died at once, their lungs destroyed by the gas. And through the night and the following day, more ashfall buried the town, metres deep. The countryside around lay under ash, lost to farming for a century. The town itself was forgotten for the next 1,500 years. It had been called Pompeii.

June 1783, and sparkly lights, like fireworks 1,400 m high, light up the horizon above the Laki Fissure in Iceland.

Downwind, a tinkling sound as glassy drops of lava return to earth: the so-called Pele's tears (named after a Hawaiian fire goddess). Closer to the fissures, the even more aptly-named cow-pat bombs splot down to the ground. Lava rises in black lakes along its 25 km length: lava that moves in slow eddies, and gleams red wherever the surface moves apart.

Eventually it reaches the rim, and spills over in a red-hot (or strictly, yellow-hot) river. It cascades over small cliffs in fiery waterfalls. As it loses heat it slows, becomes sludgy. A similar basalt flow on the Isle of Mull engulfed an entire tree without knocking it over, then cooled to preserve the tree's shape and even some of its carbonated timber for 50 million years.

The Laki Fissure erupted for eight months. The basalt lava advanced at three metres per minute, five kilometres per day: fourteen cubic kilometres of rubble riding on the tide of molten lava deep inside. Eventually it would bury over 500 square kilometres of land, including two churches and 14 farmsteads. But that was the least of it.

Fluorine tablets promote the growth of healthy teeth. But Icelandic sheep were being administered a huge fluorine overdose, and started to grow strange, devilish fangs. The Industrial Revolution was just starting to insert extra carbon dioxide into the atmosphere, and the Laki eruption supplied a noticeable upward blip on top of that. It also emitted sulphur oxides: over those eight months the acid rain from the fissure managed to match the output of the world's polluting industries today.

A dry fog of sulphuric acid hung over Europe and part of North America; the equivalent of the killer smogs that hit London in the mid-twentieth century. The sun was stained red, as if it had been soaked in blood. Parish records suggest that in Britain, the sulphur smog caused 20,000 extra deaths, as well as a lightning fireball in Hampshire that struck down three people. Mostly it was traced back to God himself – humans of 1723 being just as sinful and worthy of being struck down as in any other year. But Benjamin Franklin, then US Ambassador to Paris, correctly attributed the strange fog to Iceland's eruptions. Harvests failed all over Scandinavia, and in Iceland itself three quarters of the livestock starved to death, along with one quarter of the population.

LEFT The fossil tree, embedded in a basalt lava flow, at Loch Scridain, Mull

ABOVE Basalt sea cliffs, Isle of Mull. The sea cave is a 'raised beach' feature, above the present-day high tide mark.

RIGHT Basalt at St Abbs Head, coloured brick red by weathering in a tropical climate. Clockwise from top left: vesicles, gas-bubble holes, refilled with migrating calcite; slow-cooled magma with crystals in it; block tuff, a mix of ash and pumice fragments; tuffite, volcanic ash falling into water.

It was a stable high pressure area, as occurred in the Eyjafjalljökull eruption in Iceland in May, 2010, that sent the poison gas swirling over Europe. Presumably Laki also sent up a haze of magma: droplets that cooled into fragments of glass, but without disrupting anybody's holiday air travel as air travel had yet to be invented.

Two quite different styles of eruption. Pompeii, sudden and spectacular. Laki even more lethal, but in a slow and subtle fashion: and not even involving a proper volcano. The difference of style reflects a difference of substance. The Laki Fissure lies above the mid-Atlantic ridge, and the lava was (and is) from the heavy, dark, iron-rich oceanic crust. Pompeii lay (and lies) above the subduction zone where the Mediterranean sea floor is diving in under the edge of Europe. Its grey ash and pumice are formed out of light-coloured, silica-rich, continental crust.

No small drawing in geology carries more useful explanation than the subduction diagram. The theory of plate tectonics accounts for earthquakes, ocean trenches and the mid-ocean ridge, the distribution of garden snails and of mountain ranges. When it comes to mountain ranges

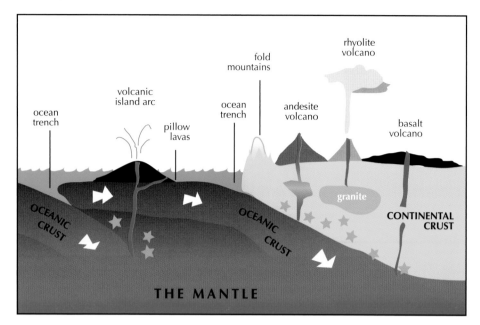

The orange stars are heat generated by the friction of the scraping plates. On the left, the ocean plates, and the Mantle for that matter, are made of heavy, black, iron-rich rock-melt of the basalt type. (Technically, it's peridotite, which is basalt but a bit blacker even.) Hot basalt magma melts its way upwards and emerges as a basalt-type volcano. The result is a line of black, basalty, volcanic islands, parallel to the shore of the continent. It's Japan and Asia today. It's England 400 million years ago, where the offshore islands would be crumpled against the continent, mangled, and eventually walked all over by people in big boots: the remains of those offshore volcanoes are now what we call Snowdonia and the Lake District.

the diagram explains not just why they are where they are, but how the heck there can be such things as mountains in an erosive world. A subduction diagram (above) is a 17-syllable haiku by a Japanese master; it's a motet by Bach; so much enlightenment crammed into three square inches.

On the right-hand side of the diagram, old ocean crust is diving under the edge of a large lump of continental crust – continental crust is lighter, which is why it stays on top. The ocean plate is a narrow one: a few hundred miles off shore, I've put another ocean plate diving in under the edge of the first one.

Over on the right, things are more messy. Sometimes, basalt magma makes its way right up to the surface to provide a basalt-type volcano. Elsewhere, though, the friction heat melts the continental rocks above. There, it's the granite-family continental rock that makes its way to the surface, to form a Vesuvius-type rhyolite volcano. But often enough, what happens is something in between. On its way up, basalt magma melts and mixes with continental granite magma. The resulting volcano is more vigorous than a basalt one, but less violent than the granite-rhyolite type. Its rocks are neither granite pale nor basalt black; they are dark grey.

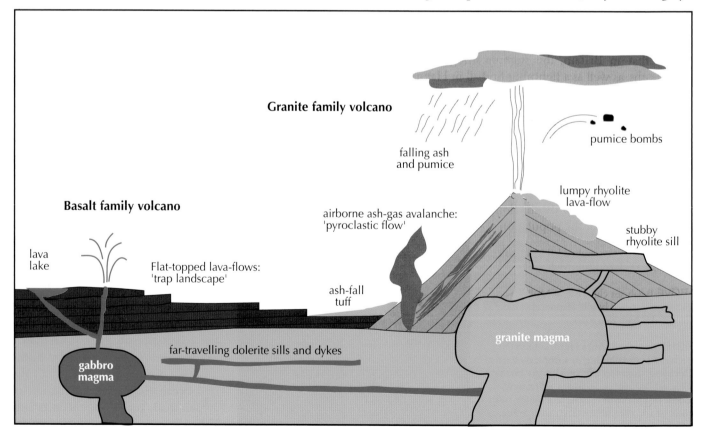

Such in-between-type dark lavas are typical of continental edges, where volcanoes erupt among fold mountains with an ocean plate diving in underneath. The longest such advancing continent edge today is the western edge of the Americas. The Andes have many such intermediate-mix volcanoes, and the resulting rock is called andesite.

Meanwhile, somewhere off the edge of the diagram on the left, two ocean plates are drawing apart, and new Mantle magma is seeping up into the gap. That's the mid-ocean ridge. Like the island arc volcanoes, the lava seeping up is black, heavy, and basically basalt.

The next diagram (far left) shows in cartoon form the two sorts of volcano. On the left is a basalt one, erupting dark, heavy, ocean-bottom rocks. It could be the Laki Fissure; it could be happening right now on Hawaii. On the right is a volcano of the granite-rhyolite family. It's erupting pale-grey lava and ash, melted down from continental rocks. If life is indeed, as Forrest Gump tells us, like a box of chocolates, then these volcanoes come in milk chocolate or in dark.

The milk powder, as it were, is the mineral silica. Silica in its crystal form is quartz, and quartz is what puts the sand into sandstone. Silica is less dense than other common planet ingredients, and tends to end up in the surface scum: or continents, as we call them. It's whitish or translucent, and lends its pale colour to the continental granite-type rocks. And as it emerges to the surface within the granite-rhyolite lava mix, there's one thing about it that really matters. In its molten form, silica is sticky.

ABOVE Gabbro is the deep-underground, crystalline member of the basalt family. Its tough rock forms jutting headlands along the north Pembrokeshire coast: this one is Penclegyr.

LEFT Seawashed pebble of Penclegyr gabbro shows its dark crystals.

Fizzy red wine that you take down to the beach has gas dissolved in it under pressure. Unscrew the little lid, and the gas fizzes out as bubbles. Underground magma also contains carbon dioxide, plus water vapour and other gasses, dissolved under tremendous pressure. Release the magma to the surface, and the gas bubbles out. If the magma belongs to the basalt family, those bubbles fizz up through the molten rock to emerge as the decorative sparkly lava fountains. But now, imagine uncorking a well-shaken canister of fizzy treacle…

Do you take your eruptions silicaceous, or plain? It's the difference between oceanic crust and continental crust; between dense, iron-rich rocks, and slightly lighter silica-rich ones. It gives you lumpy rocks that are black (or brick red); or else it gives you lumpy rocks in pale grey (or, occasionally, pink). And it's the difference between Hawaii, and Mount St Helens: between dangerous, and very, very dangerous.

On each side of the diagram, volcanic rocks occur at three different levels. Some volcanic rocks never make it to the surface. They solidify deep underground, and if we see them today it's because the deep underground has eroded off the top. These slow-cooling rocks have time to form big crystals, big enough that you can see them quite easily, and feel them as well if you happen to fall over on them.

On the basalt side, the deep-underground, big-crystal rock is gabbro. It's the stone that makes up the steep, shapely Cuillin mountains on the Isle of Skye. If you're a Skye climber you'll carry the scars of it on your knees and finger-tips. Otherwise, you may not have come across it.

On the granite side of the diagram, the big-crystal underground rock is granite itself. Granite is too well-known: as in such phrases as 'the granite crags of Everest/the Eiger/ Mont Blanc' and the black granite kitchen counter. Mont Blanc actually *is* granite, but Everest and the Eiger are limestone, and the kitchen counter is made of gabbro. If it's called granite, and it isn't white or pink with black speckles, then it's wrongly named.

At half height in the two diagrams, magma is squeezing sideways between the beds of the original scenery, forming volcanic sills. Or it's squeezing into upright cracks, forming volcanic dykes. The sticky granite magma makes big, lumpy sills that don't travel more than a few miles. The sill rock cools faster than granite, but slowly enough to form small crystals – too small to be seen without a magnifying glass. The rock is the colour of granite, with the speckles mixed back in, and it's called microgranite.

At mid-level on the left, the fluid basalt magma can travel for hundreds of miles. And when the great, smooth sheet of it cools, it cools in a regular way, developing the cracks called columnar jointing. These thin dykes and sills cool quickly. The rock is black like basalt, and its very small crystals are invisible. It may be referred to as basalt; strictly, it's the intrusive version of basalt and should be named as dolerite.

Up on the surface, basalt lava flows out eagerly, layer on layer. When those layers are level and even, the lava cools with columnar jointing, seen at its most spectacular on the Giant's Causeway or on Staffa in Scotland's Hebrides.

The dolerite sills also did columnar jointing. How do you tell which is which? Sometimes you can't. But sills have smooth tops and bottoms, while lava flows over the bumpy ground, and swirls around all irregular on top. If it then lies out in the open for a century or two, it rots in the rain and wind, giving a layer that's red and crumbly. So red topped layers are lava-flows; flat topped layers are sills.

The lava-flow form of granite is called rhyolite. It flows reluctantly, like red-hot porridge. It settles lumpily, sometimes with swirly porridge patterns inside it. Rhyolite lava is tough, chunky, and pinkish or else pale grey. Quite often, it contains speckles of feldspar, in white or grey or pink.

But rhyolite lava is just a small part of what comes out of this right-hand volcano. Volcanic ash falls to form stuff called tuff. Tuff can be featureless and grey, but is often patterned inside with lumps of pumice. Rubble slides down the volcano side. Piled up with ash it forms block tuff, a sort of crazy paving but in three dimensions.

TOP LEFT At the cliff top, a thick basalt sill, with the vertical columnar jointing that formed as it shrank and cooled. On the foreshore, a narrow basalt dyke is at bottom right in the corner of the picture and reappears in the small pinnacle. As the dyke is vertical, its columnar jointing is horizontal. The brown sandstone rocks below the big sill are Jurassic, with belemnite fossils. Bearraig Bay, Isle of Skye.

LEFT Granite-type rock that cools quickly (in lava-flow or intrusive dyke) has crystals too small to see without a lens. Because rhyolite rock is sticky (viscous), rhyolite dykes do not penetrate far from their parent granite. This one is only 100 m from the Criffel granite, Dumfriesshire coast.

ABOVE Camel's Back, a basalt dyke close to the Giant's Causeway

Art of the Novel

When in doubt, says Raymond Chandler, have a man come through a door with a gun in his hand. The whole art of the novel, in among the lyrical nature bits and the conversations, is to slip in the occasional car chase. Even in the mildest of beach-towel light reading, after five or six chapters of gossip and shopping, they'll take off their designer-labelled clothes for a quick sex scene.

From Fife to Brighton, and back up the west coast to Scotland again, the coastline is a well-ordered story of sedimentary layers. It's an intriguing story, certainly, with its complications and its fossils. But every five or ten miles, out of the sand there leaps a strange, unstratified lump of crag with, as it were, a gun in its hand. Okay, so the gun went off about 200 million years ago. But there's grim stony reality in the ruined castle standing (as often as not) on top.

These are the rocks called igneous, meaning that they're derived from fire. In a world where air and water erode away the earth, it's the hot rocks, the leaping lava and the intrusions, that prevent everything ending up as sludge at the bottom of the sea.

On the Fife coast, with its pretty red-roofed villages and subdued Coal Measure coastline, up leaps a line of blobby-looking black sea stacks: the wave-broken remnant of a big basalt intrusion. Out at sea, the Isle of May shows the flat top and columnar jointing of a dolerite sill.

PREVIOUS PAGE St Abbs Head, Berwickshire, offers 5 km of beautiful basalt. A tropical climate has attacked the iron minerals in the lava and turned the originally black rock to brick red.

LEFT Basalt lava-flows interrupt a coastline of red sandstone: Gullane, East Lothian

ABOVE AND BELOW An eruption of the Heads of Ayr volcano provided this tuff, a rock composed of volcanic ash falls, at Greenan Castle, Ayrshire.

Among the red sandstone of East Lothian, regular as brickwork, the pattern suddenly breaks into strange swirls. By chance, the colour is the same, but the texture is half-melted and strange. Someone round here's been letting off lava.

Northumberland is a succession of limestone, and shale, and patterned delta sandstones. Again we're in among those orderly and useful rocks, the Coal Measures. And then you suddenly meet, just south of the kipper village of Craster, a headland of solid dark brown, where instead of level bedding, the line markings are up-and-down. It's the Great Whin Sill, having circled half of northern England, now striding out to sea. Two miles down towards Alnwick, a black channel strikes across the strata like a vapour trail crossing out the clouds. But in this case it's nothing to do with the Whin Sill, but came along 200 million years later: it's basalt from the Mull Volcano.

On first reading, the Dorset coast is like a novel by Jane Austen, disappointingly short of volcanic violence. But *Persuasion*, if you're paying attention, does contain a thrilling sea battle. And Budleigh Salterton beach, when the salt water brings out the colours of the pebbles, contains small jewels of granite and mysterious igneous rocks. The pebbles of Yorkshire are equally pretty, but less mysterious. The red, purple and black speckled crystal lavas have arrived across the North Sea from Norway, carried either by glacier or on floating ice floes. In what is perhaps a crude literary effect, humans have scaled up the igneous imports; the sea front at Staithes carries great black lumps of larvikite, a crystalline gabbro-like rock of the basalt family.

TOP Lava bombs have fallen into volcanic ash: Vean Hole, Tintagel

RIGHT The rock could be mistaken for a conglomerate (puddingstone), a sandstone-type rock containing waterworn cobbles. But note the irregular shapes of the dark chunks included in the rock. Also, even more significantly, those darker chunks are softer than the material they're embedded in.

North Cornwall is not by Jane Austen. If it's a novel at all, it's a grim tragedy by Thomas Hardy. And among the tight lines of the black slate lurks a pile of basalt ash, untamed and shapeless, peppered with lava bombs like the cannonballs that might have been fired at Tintagel Castle.

A novel consisting entirely of gun battles and the occasional car chase? It might be worth reading; I'd have to take it down to the beach and see. Scenery consisting entirely of igneous? That does work, that's effective scenery. Few people walking over Snowdonia, or the Lake District, or Ben Nevis, will wish to vary the tone with some level-lying sediments. It just happens that the grandeur of the volcanics happens on the whole in Britain's mountains, and our northern islands.

Leaving aside the mountain islands of Mull and Skye, there's just one black-rocked basalt-family seaside: the Antrim coast of Northern Ireland, visited in the last chapter. Granite forms several of the UK's most distinctive mountain ranges: Arran, the Mountains of Mourne, the Cairngorms, Dartmoor. The Mountains of Mourne may run down to the sea, but the granite doesn't quite get to the shoreline. Again ignoring the Scottish Highlands, there's just one place of the UK mainland where the granite confronts the waves of the sea.

BELOW Trebarwith Strand, Cornwall. Where volcanic rocks are in the cliffs, a steep-sided, circular island is often offshore. It's the vent of the original volcano: refilled with the welded debris called agglomerate.

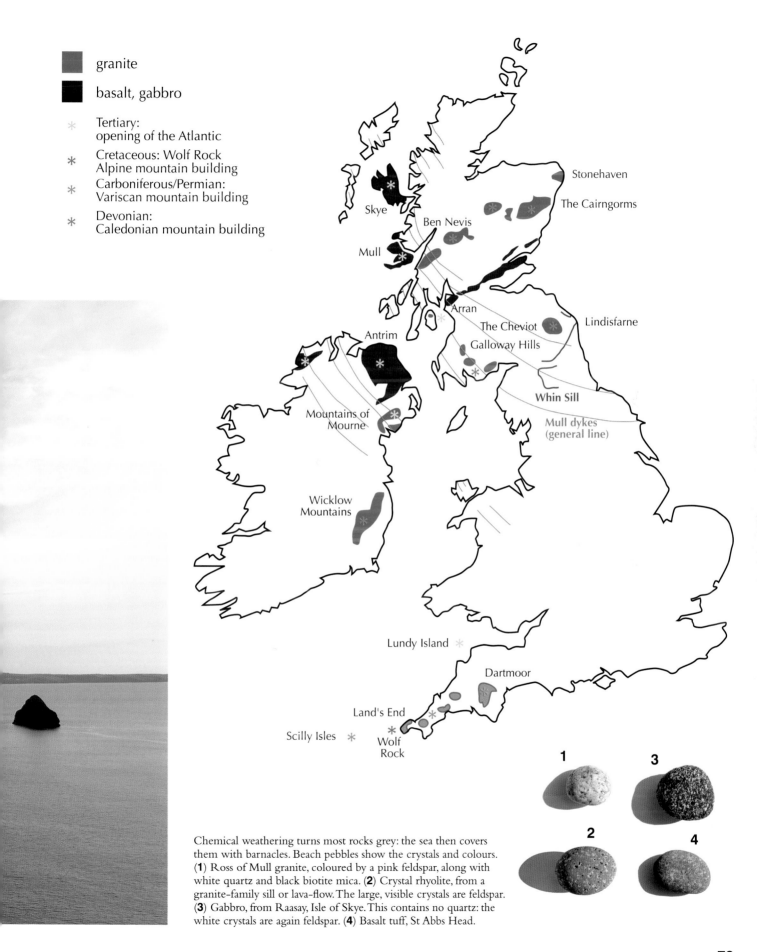

granite

basalt, gabbro

* Tertiary:
opening of the Atlantic

* Cretaceous: Wolf Rock
Alpine mountain building

* Carboniferous/Permian:
Variscan mountain building

* Devonian:
Caledonian mountain building

Stonehaven
The Cairngorms
Skye
Ben Nevis
Mull
Arran
The Cheviot
Lindisfarne
Antrim
Galloway Hills
Whin Sill
Mountains of
Mourne
Mull dykes
(general line)
Wicklow
Mountains

Lundy Island *

Dartmoor

Land's End

Scilly Isles *

Wolf
Rock

1 **3**

2 **4**

Chemical weathering turns most rocks grey: the sea then covers
them with barnacles. Beach pebbles show the crystals and colours.
(**1**) Ross of Mull granite, coloured by a pink feldspar, along with
white quartz and black biotite mica. (**2**) Crystal rhyolite, from a
granite-family sill or lava-flow. The large, visible crystals are feldspar.
(**3**) Gabbro, from Raasay, Isle of Skye. This contains no quartz: the
white crystals are again feldspar. (**4**) Basalt tuff, St Abbs Head.

79

Land's End

Just past Bristol on the motorway, they've put up signs to save you from misleading the tiresome kids with their 'are we nearly there yet?' Penzance, Truro: still more than 100 miles to go. Land's End is a long way away. And when you get there? It's a car park that genuinely is a large one, and an appropriate tourism experience to reward your long journey. There's the Dr Who phone box, recreated in fibreglass 15 metres high. There's the Land's End shopping outlet. There's a choice of fast food outlets; chose between a locally colourful Cornish pasty, or a more cosmopolitan hamburger. Land's End is a nasty, windswept spot, so all this is enclosed within a sheltering wall of inward-looking buildings.

But the natural world is present also. For £3 you can enter a darkened room, and enjoy extracts from the BBC TV *Coast* series. For those who really want to experience Land's End, this could be your spot – presumably they have the clip from Series One with Nicholas Crane exploring the wild nature 50 metres away. If you must insist on seeing that wild nature for yourself, well, there's a Land's End webcam. There's even a geological display: a lump of the Old Red Sandstone from John o' Groats, matched by a chunk of granite someone picked up from the car park.

That car park granite is, in UK coastal terms, a curiosity. The UK's only granite sea cliffs are here, at Land's End, and at Peterhead in Scotland. Otherwise, to see what the sea does to our most monumental rock – well, it's the Scilly Isles, an extension of the Cornish granite. It's Lundy, an oddity that's nothing to do with Cornwall at all but an outpost of the Mountains of Mourne. Or you would have to go all the way to Corsica.

Corsica is compelling. But so is Cape Cornwall, and the miles of rugged cliffs at Cornwall's western end, once you swerve around the tourism enclosure and past the first quarter-mile of fenced path ('dangerous cliffs' – of course dangerous cliffs, why else d'you think we came?). Land's End lacks the layered elegance of the Lias. Neither is it the black angular confusion of North Cornwall's slates, South Scotland's greywacke, or Pembrokeshire. Granite is, chemically, the simplest of all stones. It's a basic rock melt, solidifying several miles below the surface and exposed, millions of years later, by erosion.

All food, from the Cornish pasty to the finest French dining, boils down originally to a peasant diet of porridge. The rock equivalent of porridge is granite. Grind it down into sand, and you can remake it as sandstone. Reheat it and stir with a big spoon, and you'll get all the variety of gneiss and schist. Or simply pour it out, allow to cool, and serve, lightly garnished with some sea salt. Granite even looks like porridge …

Rocks that cool deep underground have time to grow crystals. Granite is a simple mix of three crystals: shiny quartz, white or pink feldspar, and little black bits, which can be either biotite mica or amphibole. The feldspar and the black bits are not tough chemicals. Out in the rain and wind, they break down into clay. Quartz, on the other hand, is very tough stuff indeed. Lacking a diamond, you can carve love-poetry into your beloved's bedroom window with a chunk of quartz. So granite, left out in the open, weathers down with rounded corners, and drops quartz crystals on the ground to make a chunky off-white gravel. It looks soft, like the butter that you left out and the cat's been licking. But brush against it in your shorts, and the sticky-out quartz crystals give you a bloody leg.

As it cooled underground, the granite shrank. As it shrank, it cracked apart in a rectangular pattern. Granite cliffs are piled blocks, like the ruins of giant castles. And because of the crystals, those blocks are rounded at their corners. Huge, rectangular, piled blocks: but let the sea attack them below, the salty winds above, let gorse and heather grow a tapestry carpet out of those thin soils of quartz gravel, and the landscape is that of Land's End.

RIGHT Land's End granite. The rectangular 'Cyclopean masonry' effect formed as the granite lump cooled and shrank, and is more recently emphasised by rain and frost.

BELOW At some point in its underground cooling, the Land's End granite happened to linger at just the right temperature for its feldspar to form these finger-sized crystals.

5. HOW TO MAKE MOUNTAINS

Folded Old Red Sandstone, Little Castle Point, Pembrokeshire. The earliest theories of the Earth would have these rocks being raised from underneath by expanding granite (Plutonism) or laid on an irregular surface by a shrinking ocean (Neptunism). The look of this sea stack suggests that they've actually been squashed sideways.

5. HOW TO MAKE MOUNTAINS

In seventeenth century Berwickshire, James Hutton realised that sandstone gets eroded away by rain, and forms sand: that sand gets washed into the sea, and forms sandstone. That left two big questions. Where did the start-off rocks come from? And, given that sandstone-to-sand sends stuff generally downhill, what gets any of the stuff back up again?

Hutton worked out that the start-off rocks were hot ones, from below the surface of the earth. And he believed that his hot new idea also solved the second question. Lumps of granite, surging up all molten from underneath, shoved the overlying rocks upwards into mountains. This does indeed happen. But it's not enough to raise the Alps and the Himalayas. And it's not enough to cause the crumples and ferocious folds we see in the coastline of north Cornwall, Pembrokeshire or Southern Scotland.

But back in the seventeenth century, they knew exactly where those folds and faultlines had come from. They were the swirling about and mountain-making caused by Noah's Flood. Their evidence for that was in the Bible. Today, we believe that it's all down to continents crashing into one another, and moving tectonic plates. Our evidence for that? Well, if the way that mountain building (along with the opening of the Atlantic) has configured the UK means anything, it means that it's quite often raining. While the drops rattle on the tin roof of the beach hut, and waves thrash the wet shingle, it's a good time to light up the gas heater, shake the sand out of the sofa, and take a survey of our evidence for this implausible idea.

LEFT Extravagantly folded greywacke sandstone, Monreith, Galloway

RIGHT An upward fold is called an anticline (think A for arch) whereas a downward one is a syncline (think S for sink). This is an unusually tight anticline at Monreith.

BOTTOM LEFT The rock has not only been folded, but also broken and moved past itself – a fault.

TOP RIGHT The wave-cut platform gives a horizontal slice through a vertical fold. St Monans, Fife

1. Africa snuggles up to America

A Frenchman called François Placet first spotted it in 1668. Using the not-that-inaccurate maps developed in the Age of Exploration, he pointed out that the western edge of Africa matches with the east edge of Brazil, the corner tucking in neatly to the great Gulf of Guinea. Clearly the two continents had once been joined together before breaking apart. He placed that breakup way, way back in the depths of geological time. As early, indeed, as 5,000 years ago, at the time of Noah's Flood.

This idea was so absurd that it was sensibly ignored for the next couple of centuries.

By the 1850s, imaginative geologists were regretting more and more that it was impossible for continents to wander across the earth. The wonderfully-named Antonio de Snider-Pellegrini drew a map matching all of South and North America against the coast of Africa and Europe. Not only did the Coal Measures of England match up with those of Pennsylvania, so also did the various fossil ferns and horsetails within the coal. As a modern Frenchman, Snider-P had moved on from the 'Noah's Flood' theory of continental separation. The Atlantic actually opened up following a volcanic explosion on Day Six of the Creation (Saturday 30 October 4004 BC). The punishments of the Flood only applied to the Old World. The so-called American Indians descend from Adam directly, rather than via any offspring of Noah, so God didn't need to drown them.

Fifty years later, the continental shelves were discovered and mapped during surveys for undersea telegraph cables. Interestingly, using this 'real' edge, Africa matched America even better than before. In 1910 American Frank B Taylor pointed out how continental wandering could account for the world's main mountain ranges. If the Americas had indeed shoved their way westwards away from Africa, then the crumpling of the leading edge would conveniently create the Andes and Rocky Mountains. Meanwhile Africa and its small sidekick India, moving away in the other direction, would crash into Europe and Asia, and thus make all the ranges from the Pyrenees via the Alps, Carpathians and Caucasus right across to the Himalayas.

But when and how did this happen? The Book of Genesis wouldn't do any longer, and a completely new wrong theory was called for. That theory (devised by Charles Darwin's less inspired son George) involved the Moon surging up out of the Pacific Ocean under the attraction of a passing star.

Tropical coral at Borron Point, north Solway coast – where the sea-bathing today is not noticeably like the Caribbean

2. Tropical reefs in England: ice in Brazil

My oldest geology book, which dates from 1943, gives a confusing story of Britain. Hot desert conditions prevailed, followed by tropical jungle and coral reefs, followed by more hot desert. For a while we were volcanic, and then a warm sea washed over us, but flip over a dozen pages and crikey, we're in an ice age.

Did the world's climate work quite differently in the Long Ago? Or else, was England in an altogether different part of the world? There could be a third possibility, that the World itself flipped over, swapping round poles and equator. That, however, turns out to involve rewriting Standard Grade physics. It takes a truly enormous force to realign a rotating Earth. A passing black hole could do it, but not without breaking the world into bits. It is slightly easier to imagine just the earth's outer layer sliding about, like the skin on an over-ripe peach.

The idea that the climate may have been altogether different: that one might be the first choice. The recent (or, strictly speaking, current) Ice Age happened with England in its present position. So why not, much earlier in our Earth, a warm and sunny 'Nice Age'? The closer you look, the less likely that one becomes. In the Carboniferous period, tropical coral reefs flourished on the coast of Wales and the Isle of Wight. They flourished at Borron Point, in Dumfriesshire, which is often a chilly spot today. They were also in eastern Canada and – good gracious! – in Greenland. Meanwhile, looking closely at the rest of the world, central Africa and Brazil were suffering an ice age.

The man who looked closer was called Alfred Wegener. Wegener was an astronomer and meteorologist – so whenever it was too cloudy to look at the stars, he could just look at the clouds. In his spare time he enjoyed exploring Greenland. But in 1911 he glanced into a geology book that traced the 1500-mile land bridge and the lost continent of Lemuria, and showed how that explained the correspondence of seed-fern fossils in Madagascar and Southern India. And he went: 'Oh! come now.'

Wegener tracked the continents back, matching up the shapes of their shorelines, the fossils found in their ground, the mountain ranges, and the geology. Closing the Atlantic to bring South America next to Africa was just the start. He moved the continents earlier and earlier, through the Cretaceous, the Jurassic, right back to the Carboniferous and the Devonian. He was able to extrapolate, at the start of the Permian period, a single, huge continent, which he called Pangaea, containing almost all the current land of the entire Earth. The jigsaw matching accounted for the coalfields spread across Britain and North America. It connected

together the New Red Sandstone into a single desert, a quarter of the world wide, filling the rainless inland area of that huge continent. During the Carboniferous, Wegener's moving map not only had England in the tropics, it had South America and Africa passing slowly and with great dignity across the South Pole.

Glossopteris indica was an 8 m-high seed fern that flourished in the Permian period. Along with its close relatives it made up the 'Glossopteris fauna' in India: but also in southern America and southern and central Africa. Glossopterids made up over half the leafy plantlife of Australia, lasting right up until the Great Dying at the start of the Triassic when the whole family got finished off (thus providing a great opportunity for the conifers).

That distribution had already called for some long and unlikely land bridges across the Atlantic and Indian Ocean when Captain Scott and his party found the first Antarctic *Glossopteris* beside the Beardmore Glacier. Even with death so obviously ahead, the expedition did not abandon the 16 kg of stones that showed fertile forests underneath the ice. They were found on the sledges alongside the bodies of the expedition members. By an odd chance, in 1930 Alfred Wegener was to die in the same way on the Greenland icecap.

Perhaps Wegener's finest contribution was in refraining from offering up any wrong theory as to how continental drift could happen. He simply displayed all the evidence that it had. Even so, the theory didn't go down with the fossil folk, geophysicists, oceanographers and geologists whose professional toes Wegener had trodden on with the entire weight of the Earth behind him.

Fossil folk were well aware of the corresponding fossils of Newfoundland and England, but supposed a land bridge now sunk under the Atlantic. Meanwhile, Wegener (who was a weather forecaster) had got one or two things wrong about fossils. The geophysicists knew perfectly well that Atlantis has never existed and there is no sunken continent under the sea. But they supposed the fossil matchups were just coincidence. Meanwhile Wegener had got one or two things wrong about geophysics.

But as well as a weatherman, Wegener was an astronomer. By measuring star positions, he had himself established that Greenland was, right now in the present day, on the move at 36 metres per year. This most convincing bit of evidence was, unfortunately, one of the things Wegener got wrong. His astronomy was incorrect by a factor of 4,000. The actual annual travel of Greenland is just 2 cm.

Worst of all, just after the First World War, the man was a German.

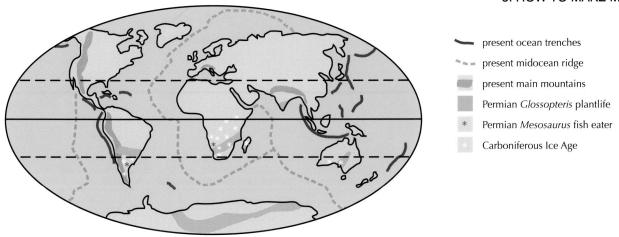

— present ocean trenches

- - - present midocean ridge

present main mountains

Permian *Glossopteris* plantlife

* Permian *Mesosaurus* fish eater

* Carboniferous Ice Age

3. Polar Wanderers

> Some circumstantial evidence is very strong, as when you find a trout in the milk.
>
> Henry Thoreau (Journal, 1850)

FAR LEFT Polar wanderer: the last photo of Alfred Wegener (left) in 1930. He died on the Greenland icecap a few weeks later.

ABOVE See plate tectonic theory making sense of the fossils and ancient icecap in the maps at the end of the book.

BELOW Earth's magnetic field

There's Relativity, and there's Quantum Mechanics. Just after the Second World War, the English physicist PSM Blackett was combining magnetism and gravity into the third great physics brilliancy of the twentieth century. Or it would have been, if it had turned out to be true. (No need to feel sorry for Blackett: he'd already won the Nobel Prize for helping discover the positron and the meson.) In the course of the physics brilliancy he'd adapted magnetometers used for detecting German submarines to make them even more sensitive. When the physics theory failed, he was left with that apparatus for measuring extremely small magnetic fields.

When he invited the geologists to point his magnetometers at the earth's rocks, they made two discoveries. The first of them confirmed, in an utterly unexpected way, the fact of continental drift. The second gave the explanation of how it could possibly be happening.

The first discovery came when the magnetometer was deployed on land. Like most advanced physics, it requires a couple of diagrams to explain it. Iron has several oxides, and one of them, with 3 atoms of iron to every 4 of oxygen, has the property of being magnetisable, as raw iron itself is.

Consequently, it's called magnetite. It's black in colour, and a good crystal is the shape of an octahedron. Magnetite isn't a very common mineral, but it is found in the blacker and more basaltic sorts of lava. (It's also found as very occasional pebbles on Chesil Beach, Dorset – but these have been traced to the wreck of an iron-ore ship in 1914.)

When lava solidifies, magnetite crystals get slightly magnetised north-south by the earth's own magnetic field. And if that lava gets broken down and washed away to make sediments, the enclosed magnetite crystals once again line themselves up north-south. Then, as the continents slide around the world and turn, the rocks retain a faint magnetic memory of where north used to be, back when they first formed.

The earth's magnetic pole isn't in northern Canada at the point on your globe marked 'North Magnetic Pole' – but somewhere deep underneath. In the northern hemisphere, a compass needle points slightly downwards – or would do, if it wasn't counterweighted to make it stay horizontal. At that point marked 'Pole' near Ellesmere Island, the needle would point directly down.

If you measure not only the direction of the magnetite's remembered north, but also its downward tilt, then you find out how far away that magnetic pole was. It records, both direction and distance: ergo, the actual position of the pole.

On any one continent, all rocks of a given age give the same position for where they think the magnetic pole is. But since then, the continent and its rocks have moved, so that position has little to do with where north actually is today. Older rocks will put the pole somewhere else again. For any continent, you can draw a line on the map tracing where it remembers the north magnetic pole as being at various points in the past. It's called a polar wandering curve; which is a bit of a misnomer. The pole does wander about

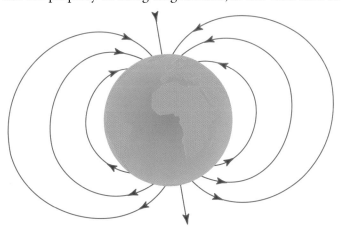

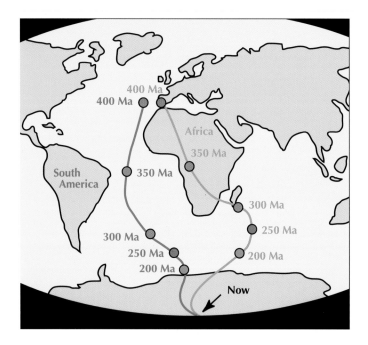

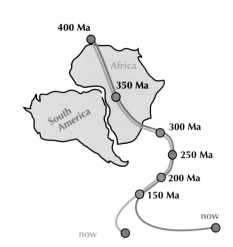

- ● Devonian
- ● Carboniferous
- ● Permian
- ● Jurassic

ABOVE Iron rich rocks, as they solidify, preserve a memory of the Earth's magnetic field. The diagram shows where rocks in Africa, and those in South America, remember the magnetic pole as having been.

Note that the magnetic pole was never actually at Gibraltar. The magnetic pole was actually somewhere close to where it is at the moment: it's Africa and South America that have moved, not the magnetic poles.

TOP RIGHT We can move South America, and its polar wandering curve, so as to make its curve match exactly with Africa's for the period between 400 Ma and about 150 Ma. (To make this work right it won't do to take scissors to the first diagram: you have to plot it all on a curved globe.) And when you do this, S America nestles up against Africa exactly as predicted by the theory of continental drift, with geology, fossils and coastline all matching along with the polar curve. Conclusion: the two continents were glued together until 150 Ma, at which point they started to move apart to their present positions.

BELOW The black magnetite crystal in this Scottish pebble is 2mm across, which is just big enough to deflect a compass-needle if held really close.

a bit (magnetic north has shifted by seven degrees since I started walking the hills), but what the curve is really recording is the wandering of the continent.

On the diagram are the polar wandering curves for Africa and for America, for the last 400 million years – back to the Carboniferous, when we had all those tropical corals and Africa had its ice age. Measure the position of that re-membered magnetic pole, at various times, and try not to be thrown out by rocks that have shifted since they solidified.

The north and south magnetic poles do occasionally switch off and restart the opposite way around. It's easy to allow for this, and the diagram here plots the wandering of what is currently the magnetic south pole. Accordingly, at year zero, otherwise known as now, both curves have the magnetic pole where it actually is, down at the bottom

of Antarctica. At 300 million years ago (300 Ma), South America's ghostly pole is in a particularly empty bit of the South Atlantic near Tristan da Cunha. Africa's one is just south of Madagascar.

Two magnetic south poles? Not at all. This indicates that South America, or Africa, or more plausibly both, have them-selves been moving about. Two hundred and fifty million years ago, South America's remembered pole is just off Brazil. If you put the south pole back where it belongs, that places Brazil in the far frozen South. This fits well with the ice-scratched rocks and debris of South America's Carboniferous period.

But it gets better when we compare the two curves. Both of the curves necessarily end up with the South Pole where it is today. But going back over the last 300 million years, the curves move apart. This must mean relative move-ment of South America vis-à-vis Africa. That relative move-ment could be either towards, or away-from. It's going to make more sense as away-from, from the opening up of the Atlantic. From 250 million years ago to now, the polar wandering curves converge on each other, but this actually marks the two continents moving apart.

Between 400 Ma and 250 Ma, the two curves look rather alike. If we drew them on a curved globe, they would look exactly alike. Now move the South American curve until it lies exactly on top of the Africa one, and at the same time move America itself. We find we've just closed up the South Atlantic, and combined Africa and South America into a single landmass. The diamond mines of Brazil meet the diamond mines of South Africa. The Mesosaurus, a spiny-toothed lizard-like eater of fish, lives in a single set of rivers. And at 350 Ma, the South Pole is right in the middle, nicely explaining those glaciers in the Carboniferous. At 150 Ma, the two curves stop overlapping. That, then, is when the con-tinent cracks open, and the South Atlantic begins to open up. To confirm this date for the start of the South Atlantic, seek out the earliest mid-ocean ridge basalt.

Incredible shrinking Earth

Imagine one of those grim seaside boarding houses that, quite possibly, never existed even in the dark days of the 1950s. You come in all sandy at half past four, and in the gloom behind the dusty lace curtains, on the small bendylegged table, mid-positioned on the lacy doily, there's a teapot. You touch it; it's slightly warm.

Seeing as it's warm, you know without being told that it's cooling. To put it in a fussy way, it's losing more heat to the air around it than it's receiving back from the dank unheated dining room. It's cooling particularly quickly as the churlish landlady left off the tea-cosy. Given that it's cooling, you can make a guess at how hot it has been previously; and estimate that you should have come in for your teatime at round about half past three. Probably not worth asking Mrs Thing to bring back the biscuits …

The Earth, too, is a teapot – one of those homely brown ones, made (obviously) of earthenware. The Earth, if you carefully calculate all the gains and losses, turns out to be giving off slightly more heat than we receive from the Sun. In other words, the Earth is currently cooling. A five-cup teapot with no teacosy cools in about an hour. How long till the Earth gets tepid? William Thomson (later Lord Kelvin) worked it out in 1862 and got the answer: fill up the teapot with redhot melted rock, and let it cool for about 20 million years.

Some geologists thought this was a bit short for all the exciting events of the 11 geological periods. But the teapot is basic physics, and the figures have a lot of leeway. Up to the 1850s, the cooling earth explained the changes in the fossils – at least until Darwin's theory of evolution offered an even better explanation.

Anything that cools down, also gets smaller. Concrete road bridges have rollers under one end to allow for hot summers and cold winters: you can examine them from your beach-bound Bank Holiday traffic jam on the M4. And an Earth that cools is also an Earth that shrinks. If the Earth.nterior cools by, say 500° C, the shrinkage is about 2 per cent (by volume) or a bit less than 1 per cent in diameter. And as the Earth shrinks, the crust wrinkles up. There you go: mountain ranges!

But also, as the Earth shrinks further, the wrinkles could perhaps rearrange themselves. What was wrinkled down could later be wrinkled up; so that sea-floor sediments could end up high in the mountains. And what was wrinkled up could later be wrinkled down, and this was even more important. Where did they go, the 'land bridges' that allowed Mesosaurus, the spiny-toothed water lizard from Central Africa, to wander across into South America, or delicate splay-leaved *Glossopteris* seed ferns to spread from Madagascar all the way to southern India? Conveniently, the shrinking Earth's wrinkle-rearrangement lets the land bridges sink gracefully back beneath the sea.

What's wrong with all this? The Earth is indeed slightly too hot: but this is not because it hasn't quite finished cooling. The Earth, it turned out at the start of the twentieth century, has a heat source inside it. That heat source is the radioactive decay of uranium and other elements down in the Earth's core. It's as if that unfriendly landlady had placed a very small electric element inside the teapot so that, *whatever time you came in for your tea*, you would still find the teapot slightly tepid.

4. Sea-floor Spreading

The next stage was to take Blackett's gadget back to where it had started – the Atlantic Ocean. Instead of looking for enemy submarines, it was to detect the much weaker magnetism of the ocean bed beneath. Straight away, they found that the deep ocean floor was much more magnetic than the land and the shallow sea. The continental shelf, as already suspected, does indeed mark the edge of something.

But as well as being stronger, as you sailed across the Atlantic, the ocean floor's magnetism switched from north to south and back again.

Tracking the polar wandering curves of the continents had already shown that the earth's magnetism does reverse north to south from time to time. So this back and forth magnetism makes sense if different bits of sea bed solidified at different times. Now if you colour 'north points north' parts of the Atlantic black, and 'north points south' parts white, you get a particular pattern: a set of stripes running roughly north and south. It's like the up-down magnetism on a strip of recording tape. And as you cross the great Mid-Atlantic Ridge, something else happens. The pattern reverses, and unplays itself backwards all the way to America. It's as if you played some pop classic through the first half of the journey, then played it in reverse the rest of the way. Like running the Beatles' *Sergeant Pepper's Lonely Hearts Club Band* through the cassette player upside down a few times (the year was 1967) – suddenly it made sense.

The continents weren't ploughing their ways across the ocean bed. They were riding on top of it, while the ocean bed itself was spreading outwards in both directions. At the mid-ocean ridge, new ocean floor was rising from the earth's molten interior, solidifying, and incidentally recording where north happened to be just now. This implied that ocean floor had to be disappearing again, somewhere at the other end. The deep ocean trenches, around the edges of the Pacific, were just the places for that to be happening.

Thus, after 250 years and to a soundtrack of *Day in the Life* played backwards, James Hutton's tricky question was answered. How do the sea-bottom sediments get back up onto dry land? The answer, all along the eastern seaboard of the Americas, is that disappearing ocean bed wrinkles up the front edge of the advancing continent and makes mountains. And the other answer is, from the Alps along to the Himalayas, that continents collide and crumple, and make even bigger mountains.

Everest was once the ocean floor in the gap between Asia and an advancing India. Seashells on its summit are only to be expected. The Atlantic is getting wider – the movement can now be measured using GPS – and you must expect the price of a first class ticket to New York to increase, inexorably, by 1p every 300 years.

Crunch times

The section opened with the tropical corals at Borron Point. Was the world a whole lot hotter 400 million years ago? Was the coral tougher stuff, adapted to our northern climate? Or has England itself been moving about a bit? The third alternative, apparently the least likely, turns out to be the true one.

To make possible the Antrim chapter above, a volcanic episode poured black basalt all over the chalk of Northern Ireland. We can now put a name on that volcanic episode; it was the opening of the Atlantic.

Weird and mysterious black rocks are on the seashore of Ayrshire, and some more of them at the Lizard Point. Plate tectonics – the up-to-date name for Continental Drift – makes them less mysterious, and at the same time even more weird.

And at various points around the coast, you see the slow-motion car crash effects of two continents colliding. The average car driver has a lifetime one-in-three chance of a bad accident. The UK, as it trundles around the globe, has been unlucky: we have suffered three.

The Alpine mountain-building, the third and last of our three crunches, is happening right now. The Mediterranean is an ocean that's closing: in less than 50 million years it will close up completely. As Africa trundles onwards into the underbelly of Europe, Holland and Belgium will become mountain lands whose mightily-lunged footballers will easily take on the South Americans. Even now, scraps of continental crust are crashing into mainland Europe: notable among them, Italy. The south coast of England feels the distant knock-on effects, as the chalk tilts upwards, and the inland domes itself gently into Cotswolds and Chilterns.

At the end of the Carboniferous 200 million years earlier, we were involved in a much bigger pile-up. Two great continents were coming together to form Pangaea, the world continent identified by Alfred Wegener. The southern continent, Gondwana, combined what would eventually become South America, Africa, India, Antarctica and Australia. The northern part had as its central feature the great desert of the Old Red Sandstone, and has been called the Old Red Sandstone Continent, but today is more commonly named as Laurasia. Its eventual breakup has formed North America, Europe and Asia.

ABOVE Mica schist in Central Park, New York, resembles the standard rock of the Scottish Highlands; the Appalachians as a whole match the Highlands in age as well as in rock types.

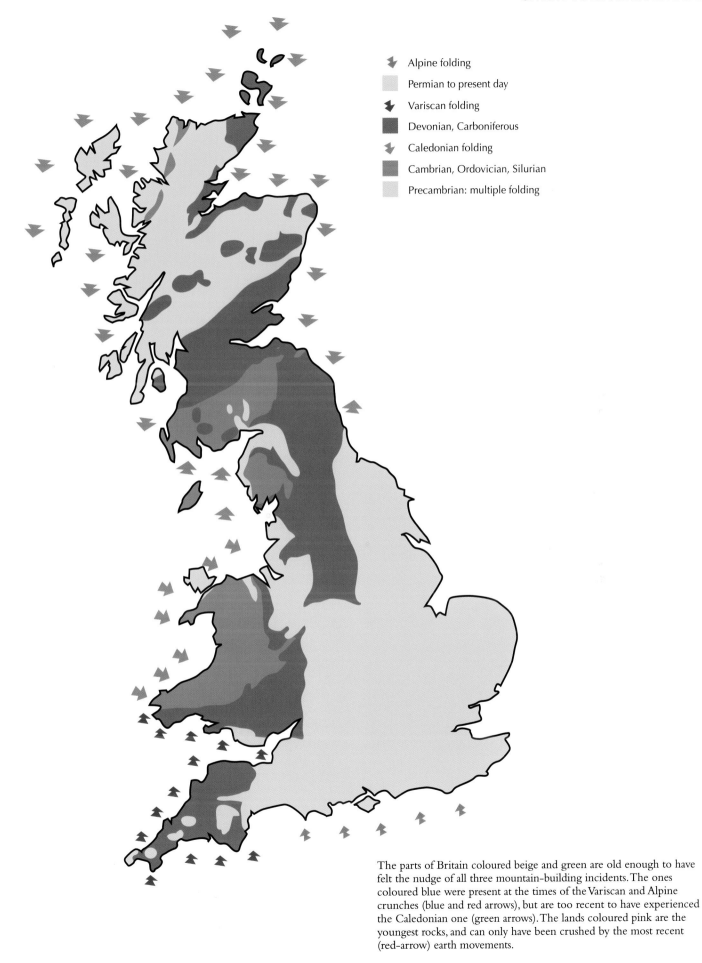

Alpine folding

Permian to present day

Variscan folding

Devonian, Carboniferous

Caledonian folding

Cambrian, Ordovician, Silurian

Precambrian: multiple folding

The parts of Britain coloured beige and green are old enough to have felt the nudge of all three mountain-building incidents. The ones coloured blue were present at the times of the Variscan and Alpine crunches (blue and red arrows), but are too recent to have experienced the Caledonian one (green arrows). The lands coloured pink are the youngest rocks, and can only have been crushed by the most recent (red-arrow) earth movements.

The UK was near, but not on, the southern seaboard of Laurasia. The actual crunch point was at the tip of Spain, where it was being barged into by the northwestern corner of Africa. The mountains raised in the collision include the Urals, the Pyrenees, the Atlas, and in the US the Alleghenies. In Britain the crumple zone extends northeastwards into Cornwall, Devon, and the south coast of Wales. Inland, underground magma swelled upwards to form the granite domes of Dartmoor, Bodmin Moor, and Land's End. The crunch-time itself is named as the Variscan – or sometimes the Hyrcanian. And to watch the continents as they inexorably crunch together, just go to Crackington Haven, on the north Cornish coast.

The earliest episode was also the one involving our island the most intimately, being no less than the collision between England and Scotland. It happened at the end of the Silurian period, and is named the Caledonian collision. The great mountain range along the crumple zone stretches from Scandinavia to the Appalachians. The worn-down roots of that range form all the UK's finest current mountains: Snowdonia, the Lake District, the

Scottish Highlands, even the Southern Uplands along the England-Scotland border.

Everything in geology is a long time ago. But the end of the Silurian, 400 million years BC, is a *really* long time ago. Although the Caledonian crunch was the fiercest one, you'll see none of its effects round England's coastline. Every rock there is on England's shore is younger than the Caledonian crunch. The sea-cliffs of Wales have been crunched Caledonian-style; but it's not easy to see, as they've since been recrunched in the Variscan.

The crash damage of Scotland vs England is most visible today in the coastline at either end of the Southern Uplands: Galloway in the West, Berwickshire in the East. In Berwickshire, in the Introduction to this book, the underneath layer of Hutton's Unconformity was greywacke sandstone tipped on edge. It was the Caledonian crunch that did the tipping.

And it was Alfred Wegener, with his theory of wandering continents, who solved Hutton's unanswered question: how those sea-bottom sediments, the ancient grey and the slightly less ancient red one, got raised back up into Berwickshire for him to see at the seaside.

LEFT On the back of the rocks at Portishead, horizontal faulting, where the upper rocks have been shoved across the lower ones, leaves the scratch-marks called slickensides.

ABOVE The Caledonian crunch has folded these deep-ocean greywackes all over the place: St Abbs, eastern end of the Southern Uplands

RIGHT From St Abbs to St Bees … the corresponding point on the west coast of England shows the strata quite undisturbed. This isn't so very surprising; these New Red Sandstones were laid down 200 million years after the Caledonian Crunch.

6. CHALK

Chalk cliff west of South Landing, Flamborough Head, Yorkshire. It's chalk all the way across England from here to Brighton, with London and younger rocks to the south and east, and older rocks everywhere to the north and west.

6. CHALK

Ink is messy and paper is pricy. Victorian children learned their letters on a rectangle of slate, with a wooden frame to protect small fingers from the sharp edges. They wrote on it with a piece of chalk, and wiped it for reuse with a damp sponge.

The writing surface, then, was almost certainly Ordovician: deep ocean mudstone, compressed by the earth movements that brought England and Scotland into one landmass together, 400 million years ago. In the writing hand is a stick of fossil limestone, from the warm seas of a mere 100 million years ago. Dinosaurs decorate the primary school classrooms of today. But the child of 200 years ago learnt to write using a small lump of rock from the times of the dinosaurs.

The chalk still sometimes used in classrooms is now made not of chalk but of calcium sulphate, which is derived from the dull, dinosaur-free evaporating salt lakes of the Permian (or New Red Sandstone) period. Anyway, computers don't like chalk dust, whether authentic organic or the sulphate replacement.

True chalk is made of calcite, calcium carbonate, and so is strictly speaking just a variety of limestone. In fact, it's an extraordinarily pure sort of limestone: it is 98 per cent calcite, which is why it's so white. In the ordinary sorts of limestone, the calcite firmly cements the rock together. The Mountain Limestone is one of Britain's toughest rocks, or it would be if it didn't slowly dissolve away in rainwater. But chalk is just

calcite fossils, gently squashed. The fossils are of all sizes from the too-small-to-see right down to the submicroscopic, and even smaller than that. The finest ones were just called 'chalk dust' until electron microscopes came along in the 1960s.

'Plankton' is a general term for very small sea life. It consists of crustaceans half a millimetre long; seashells not unlike ammonites but a mere millimetre across; and the free-swimming larval stage of all the various sorts of seashells, as well as sea anemones and coral. And in the gaps between this microscopic life, there is even smaller microscopic life.

The animal plankton fed on the algae, single-celled floating plantlife drawing energy directly from the sunlight. To try and avoid being eaten, each little green plant-blob armoured itself with overlapping shelly plates, called coccoliths, two or three thousandths of a millimetre across. Slightly less tiny types of planktonic life, crustaceans called copepods, ate the algae and excreted pellets of coccoliths onto the sea bed. And when you take your blackboard duster outside and shake it, the white powder that makes you sneeze is 100-million-year-old copepod poo. (Or it would be, if your modern chalkstick was still made of actual chalk.)

One gram of chalk contains about 100 billion of the tiny plates – that's about the number of stars in our galaxy. It's about the age of the Earth, in fortnights.

Beer, in East Devon. As this Cretaceous chalk is part of the so-called Jurassic Coast, this book's Beer festival is in Chapter 8.

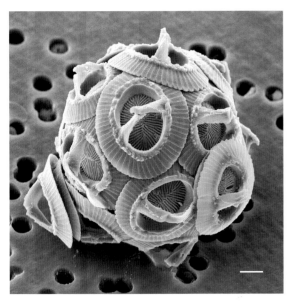

Chalk Talk

There are two sorts of stone that everybody knows without a handbook. One of them is coal, and the other is chalk. Chalk is crumbly and dusty; it forms high white cliffs with stripy lighthouses underneath; and it is traditionally quite different from cheese.

Cheese comes in whitish or yellowish homogeneous blocks: cut the stuff anywhere and it's much the same. Chalk, on the other hand – well, it comes in whitish or yellowish homogeneous blocks. Cheese comes about when you take some milky goo and squeeze out most of the water, using a heavy press over several weeks. Chalk is quite different: the heavy pressure has to be applied over several million years.

This, of course, is the point of the proverb. In fact, chalk and cheese are rather like each other. Don't the high cliffs along most of Kent and Sussex, sliced off so clean and vertical, remind you of the edge of the Cheddar, a big block in the delicatessen, cut through with the cheese wire? But sprinkle chalk on your meatballs, or write on a blackboard with a piece of ripe Camembert – then you'll perceive the difference.

If the cheese-sliced sea cliffs at Beachy Head and the Seven Sisters show one thing, it is this. A fine, exciting seacliff has nothing to do with the hardness of the rocks. The rocks of southwest Wales are ancient and tough, but they don't form cliffs like the crumbly chalk of Kent. The most upstanding cliffs of Pembrokeshire, which are at Druidston Haven, are in fact in the recent, and softer, Coal Measures. The tallest cliffs in Britain, on Hoy in the Orkney Isles, are carved out of the Old Red Sandstone, another soft and easily-eroded rock.

The softness of these rocks is, in fact, the first reason in favour of their fine cliffs. It's true that a high vertical chalk face will crumble away under the wind and rain. But even more quickly, the sea is carving into the cliff foot. No sooner does the top of the cliff fall off than it's got a new bottom underneath. If the sea defences around the base of Sussex and Kent should remain in place for the next few tens of thousands of years, it's then that we may see these cliffs crumbling back into stony slopes.

The next fine-cliff factor is the level bedding. The ancient rocks of Pembrokeshire are tilted, and twisted, and cleaved into slates, and broken by faults. Those fault shatter zones and slates may well be stronger than Beachy Head chalk in its undamaged state. But while the sea is slowly nibbling the tough cliff base of St David's Head, there are plenty of millennia for water to get into those cracks, and – a millimetre at a time – force those rocks apart. One of the various folds or faultlines will slant towards the sea, to dump off the rock load into the water.

Chalk, on the other hand, has all the strength of a pile of wet newspaper. And wet newspaper, lying flat and well supported at the bottom, stands in the rain for quite a long time before the recycling van comes along.

Brighton's chalk has been raised to its present level by the Alpine mountain building. But the Alps are quite a long way away. The chalk has been raised gently, and still lies level. A chalk layer with a slight clay content, two thirds of the way up: it doesn't have to be hard, it just has to be slightly less soft than what's below, to stiffen the cliff and keep things straight up and down.

As chalk is a sort of limestone, it dissolves in water. This makes it even easier for the sea to get into the chalk: it can carve it, and knock it about, but it can also wash it away. More importantly, chalk is porous. The cliffs of Pembrokeshire are

98

broken up by water getting into the cracks, and freezing. The rainwater that falls on chalk just runs right through it.

Chalk is almost useless for building houses; you can't climb on it, and there are less dusty, and squeak-free, ways of writing on blackboards. The one thing it's really good for is making a cracking sea cliff.

The chalk runs north across the UK landscape, from Brighton and from Beer in Devon and through the Dorset Downs; northwards through East Anglia to Hunstanton in Norfolk, and around the Wash to the Wolds of Lincolnshire and of Yorkshire.

But there's more – or at least there was. Chalk once covered all the UK except the mountain ground. Today, there's no chalk left in Scotland – or almost none. One house-sized chunk of chalk still exists on the Isle of Arran. It dropped into a volcano and survives, hard-baked but still recognisable, when all the rest of the Celtic chalklands were washed away into the sea. Flints I found on the shore of Fife could just be last reminders of the former chalk, but it's equally likely that some human carried them north in the rubbish in the bottom of a boat, or in the soil surrounding a pot plant.

It all joins up, and it's all the same stuff. It crosses under the Channel to France and on into the Netherlands and Denmark. And that's it. Leaving aside the odd patch of China or Texas, chalk is a white, Anglo-Saxon sort of stone. And fussy as it is about its residence qualification, it's also particular about its geological period. Starting about 100 million years ago, lime slime started to build up on the sea bed. It piled up, at 1 - 2 mm per century over 35 million years, to form a layer up to 500 m thick. And then it stopped.

And so it makes perfect sense to call the geological period from 65 back to 145 million years ago the Cretaceous, or chalk-bearing.

Sandstone, limestone, granite and basalt all happen in every geological time. So what made this particular era so chalky? What, 145 million years ago, caused chalk – and caused it all over Ireland, England, France and Sweden? And after 80 million years of copepod poo dropping to the sea bottom, what made it stop?

The first special feature of the Cretaceous was the high level of the ocean. Shallow seas covered so much of the continental crust that land was rather rare. Worldwide, one third of what is currently land area got submerged; including almost all of England and Wales, plus large parts of Scotland. The chalk limestone is so pure and white because it holds no land-based impurities – no brown sludge of life, no mineral staining of iron. No rivers carried sand and mud into the sea from nearby coasts – there were no nearby coasts.

The sedimentary rocks that mark ancient sea levels may later rise and fall as continents crumple. Continents themselves can shift up and down – sinking under ice, rising as erosion takes off the load. So it's difficult to measure the level of the ocean as a whole. Estimates of Cretaceous overall sea level vary: some as high as 400 m above where it is at the moment, some a mere 200 m up.

Two hundred metres more sea? This brings us straight back to the Neptunian geologists of the Antrim coastline, where we sneeringly dismissed the Bible story of Noah on the grounds:

where did all the extra water come from? The current global warming, if it should continue unchecked, will raise sea levels by a few metres. The total melting of all Antarctic and Greenland ice would add 70 m to the world's sea level. Thermal expansion of the ocean – as it warms up, it gets bigger – offers a few more metres. For the Cretaceous, a bigger explanation is needed.

The answer turns out to be – no surprise – plate tectonics. During the chalk times, the super-continent Pangaea was breaking up. All around the world, the ocean crust was being pulled apart at new ocean ridges. The Indian and Southern Oceans were expanding, the Atlantic was unzipping from the south. The Pacific was enlarging by 20 cm per year in the middle, though this was compensated by its subducting at the edges for an overall shrink in Pacific. The expansion of the other oceans was absorbed by the shrinking Tethys Ocean, eventually to be labelled down as a mere sea, the Mediterranean. Overall, there was a lot of upward bulging mid-ocean ridge. And, as a consequence, the ocean floor was mostly, as ocean floors go, young and fresh.

ABOVE Because Antrim's chalk has been baked by the overlying basalt lavas, it's a bit tougher, and is usable for building the harbour wall at Cairnlough. Basalt blocks add a decorative element.

FAR RIGHT North Landing, Flamborough. Being soluble in seawater, chalk forms sea arches. As so often, this one exploits a faultline. Note the wave-cut notch along the high tide line on the right.

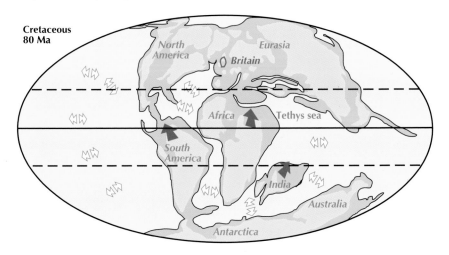

Cretaceous 80 Ma

When a tray of biscuits comes out of the oven, it stays hot for about 20 minutes. A cake stays hot for an hour. We're used to the idea that big things take longer to cool down. Geologically, a dyke of molten rock cools in a few weeks: an underground lump of granite, several miles across, cools for years and centuries, with time for its solidifying magma to form crystals.

But how long does it take a newly formed ocean crust, 10 km in thickness, to cool down? The answer turns out to be tens of millions of years. As it cools, like any cooling object it shrinks. But while it's still warm, it's bigger, big enough to shove the sea above it up by hundreds of metres.

So the chalk time was one of high sea levels. It was also hot. With the Cretaceous arrangement of the continents, the south coast of England lay further south than now, at around the north Mediterranean. But Dorset then was much warmer than Nice or Monaco today. For the Cretaceous was a time of global warming, with average global temperatures 4°C up on today. The fossil algae that make up the chalk lived in pleasantly warm sea temperatures of 20°C or more.

This conclusion starts as a guess based on similar algae today. It is verified by examining oxygen isotopes in the calcite. Oxygen 16 (ordinary oxygen) and oxygen 18 (heavy oxygen) are the same in every chemical way. But the normal and the heavy forms dissolve into seawater at different rates; and those rates depend on temperature. Pick the calcite apart atom by atom using a machine the size of a small car called a mass spectrometer, and measure the two sorts of oxygen. And then work out the interactions of rainwater, freshwater and ice to interpret how hot was the sea that the coccolith formed its calcite from.

The global warming was driven, as global warming is today, by carbon dioxide. Magma was emerging to make those new mid-ocean ridges; and all its dissolved gas was entering the atmosphere.

Cretaceous carbon dioxide stood at 1700 parts per million, or about 0.2 per cent of the atmosphere. This compares with its current level of just below 300ppm – except that human industry has just jerked that up to 380ppm. The amount of oxygen in the atmospheric above the chalk seas is estimated at 30 per cent, which is one and a half times as much as it is today. And under that out-of-balance atmosphere, over what's now the UK and western Europe, was a great, warm sea. Distant continents enclosed it: the small future Atlantic to the southwest sent no refreshing currents to stir its tepid, stagnant waters. And algae bloomed in the way they sometimes bloom in depleted lakes and reservoirs during hot summers. Today such blooms are a sign of eco-damage. The Cretaceous begins to look like a 50-million-year environmental disaster.

The Greensand Sea

The Cretaceous sea moved in westwards across the south of England. Its invading sea floor can be traced along the present-day coast between Seatown (Dorset) and Seaton (Devon), and on westwards past Sidmouth. Just as the English Channel is cutting into the land at the cliff base, and moving northwards into Dorset a foot or a metre at a time, exactly so did the invading sea move westwards, cutting into the existing country.

That sea as it moved in over the land was dirty and sandy, not clear and blue as the later chalk sea would be. That first sea floor, below the chalk but still in the Cretaceous, is the Greensand. It's an underwater sandstone, yellow where now exposed to the air, but green as its name suggests when you break it open. The green is glauconite, which is an iron mineral but one that doesn't show the cheerful red colours of oxidised iron. The presence of glauconite shows that this initial Greensand sea was already oxygen-starved and stagnant.

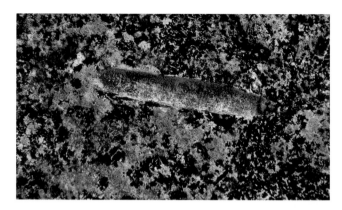

Two levels of Greensand and a clay called the Gault lie across southern England underneath the chalk. They form the lower half of the Cretaceous system; only the upper half is actual chalk. The gradual rise in the ocean made the sea invasion all the easier. The wave-cut platform of 10,000 years ago sinks out of the way, allowing the waves free progress into the next slice of cliff. The unconformity – the cut-off top of the pre-existing, older, Jurassic and Triassic rock layers – can be traced back eastwards through Hampshire to the Weald. The sea-cliff line was carved westwards right across England, as the sea sliced off the top of 200 miles of land.

The top-chopping and sea-submersion all took place during the subdivision of the Cretaceous called the 'Aptian', a timespan of a mere 13 million years. That's an average cliff retreat of 2 cm per year. This is a fairly slow sea advance. The chalk cliff at Birling Gap near Seven Sisters has retreated by 50 m since people first started taking photographs of it in 1875: that's 30 cm per year.

Above the Greensand, the chalk lies white and high. No single sea cliff shows the whole 500 m thickness of it. Occasionally there's a slightly darker band stiffened with a little clay. The darker bands help the chalk cliff, while washing away at its base, to stand up vertically above. The clay, with no local land to wash down from, may be airborne dust of distant volcanoes – Wolf Rock, off Land's End, is the plug of one of these Cretaceous-age volcanoes. One of these tougher clay layers lies at the very bottom of the chalk: it's called the Chalk Marl. Its rather more waterproof chalk carries the Channel Tunnel across into the base of the chalk strata in France.

Larger creatures than coccoliths and copepods did float or squirt themselves through the chalky seas. Ammonites were about, and sea urchins and sponges did occur. Chalk fossils are soft-edged, and can be disappointing. In theory, you might come across weird Cretaceous ammonites; uncoiled ones, and some reshaped as cones. They were once taken to be bad ammonites, degenerating towards their ultimate extinction at the end of the Cretaceous – just as effete, dormouse-eating Romans were ripe for conquest by barbarians from the energetic north.

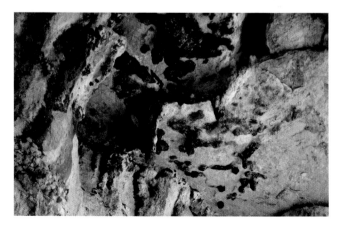

Today's evolutionary theory doesn't accept the idea of 'degenerate' species, tired out by 350 million years of natural selection. These ammonite oddities must be adapting to the unique environment of the Cretaceous sea, with its softly fluffy sea bed. In the same way certain bivalves, the rudists, evolved into tall shapes like vases, half a metre high. They may have been enjoying the coral-reef lifestyle on the delicate sludge surface.

At the very top of the Cretaceous strata, the Cretaceous to Tertiary boundary, lies a unique layer of iridium-rich clay. Iridium is a heavy element that normally occurs deep inside the earth, not at the surface. This provided the clue to the tricky question of what killed the dinosaurs, not to mention the ammonites and belemnites. It wasn't degeneration in the genes. It wasn't competition from us mammals, and it wasn't episodes of volcanism or climate change. It was, almost certainly, a 12-kilometre-wide lump of rock arriving from outer space. Its arrival made a hole in Mexico over 100

TOP LEFT Chalk is a soft rock, so fossils tend to be fuzzy-edged. Belemnite cast from Larrybane, Antrim, near the Rope Bridge.

MIDDLE LEFT AND BOTTOM Chalk cave below Larrybane (the Rope Bridge links two of the distant islands). Chalk is a form of limestone, and this cave shows rather rare 'chalkactites', chalk re-precipitated out of water dripping through a ceiling crack.

ABOVE Beachy Head from the west. The surreal effect of snow lying on chalk is rare in the twenty-first century (but the most lethal UK avalanche ever came off the chalk in 1836 to kill eight inhabitants of Lewes). The Ice-Age stream in the cliff gap was running inland. Its valley head has been completely cut away by the sea in 20,000 years since. (Photo: Fiona Barltrop)

miles across, and threw up enough dust to darken the skies of Earth for a year. Unfortunately, the crucial clay layer is not currently exposed anywhere in the UK. You'll have to visit some quarries in Italy to see it.

Flint

When you'd written your ABC on your slate with your piece of Cretaceous chalk, you needed to wipe it off again. So on a string at the corner of your chalkboard dangled a piece of sponge – originally harvested by local divers from the floor of the Mediterranean Sea. But sponges also lived in the warm chalky seas of the Cretaceous. Sponges contain a spiny skeleton of silica, the quartz material. As the chalk is being compressed from sea-bottom slime into actual rock, the silica migrates through the slime, finding and binding to other silica to form lumps – concretions – and layers. The particular sort of silica concretion found in the chalk is called flint.

Chalk is often too soft to form first-rate fossils, but flint is the opposite. Some flints form around fallen seashells, and end up enclosing the finest and most detailed of fossils. Trouble is, which flints? Flint lumps are uniquely suitable, when you didn't bother with the goggles, for flying apart under your hammer and putting your eye out.

The most exciting find, and one requiring sharp eyes rather than ones threatened by rock splinters, could be a Cretaceous sea-urchin replaced, molecule by molecule, with silica: a fossil rebuilt in flint.

LEFT Some flints form around shells, sea urchins or other fossils. Most, though, are flinty all the way through.

RIGHT Seawashed flints make a surprisingly colourful beach, but one that's uncomfortable to lie about and awkward to walk on.

BELOW Flint makes an intractable building stone, but the alternative, chalk, is too soft to be any use at all. Hotel at Beer.

Calcite

Sandstone is made of sand, and most sand is made of the tough survivor mineral called quartz. Even so, sandstones can be quite different because of the stuff that sticks them together. Geology, commonly, supplies three different sorts of glue.

Iron oxide is known to many of us as common rust. Anybody who's lain underneath an old car with a spanner knows that rust is an effective adhesive. Red iron oxide is what sticks together the standard sort of sandstone, which is red or orange, and made from desert dunes or ordinary beaches. Your old Ford Escort is mostly made of rust, but the sand particles are only lightly coated with iron minerals. So these common sandstones are friable, and break apart along bedding planes to make useful stones for building.

The sand mineral quartz can itself be dissolved in water – especially in high-pressure, super-hot underground water. When this happens, it forms the strongest sort of rock glue. Quartz is the name of the crystal: when solidified into a formless mass, the mineral is simply silica. Quartz crystals glued together with silica make the rock called quartzite, which doesn't look like a sandstone, although technically it is one. Off-white, brittle, but very tough, quartzite is quite a common mountain rock. At the seaside, its most visible appearance is as the special pebbles of Budleigh Salterton (seen on page 141).

But the commonest sort of rock glue is tougher than rust, less strong than silica, and created in the seas and oceans as a by-product of seashells. Chemists call it calcium carbonate; farmers and builders name it as lime. Considered as a rock-making mineral, it's called calcite.

Calcite forms 3 per cent of the earth's crust: either as limestone (which is calcite plus impurities) or as the cement holding together other sorts of sedimentary rock. Calcite is the lime scale that builds up in your kettle; it's the stalactites that dangle from cave roofs underground; it's the main component of seashells, and of your teeth. Pure calcite crystals are transparent: while our eyes have lenses made of water, and cameras have lenses of glass, the trilobites had eyes made of pure crystal calcite.

Calcite can be as soft as chalk or as tough as limestone; stiffened in nature's clever way with proteins it becomes the strength of seashells, and such protein-stiffened calcite is currently being investigated for aeroplanes and space ships.

Various different materials are referred to as lime, so let's clear up the chemistry. Calcium, which is chemically a strange sort of metal, usually exists on our planet's oxygen-rich surface as calcium oxide, CaO. In the sea, this combines with dissolved carbon dioxide to form calcium carbonate. This carbonate is dissolved in the seawater. But Life, whatever else it may be, is a chemical engineer. Molluscs, sponges and corals have no trouble extracting the dissolved calcium carbonate and turning it into body parts and seashells and even, in the case of those trilobites, into eyeballs.

LEFT Calcite veins often have a lumpy look to them. Vane Hole, near Tintagel, Cornwall.

TOP RIGHT Sandstone, held together three ways. Top, desert sandstone (Permian) cemented with iron oxide. Centre: sea-bottom sandstone cemented with calcite. Bottom: quartzite; quartz sand cemented with the quartz mineral silica.

This carbonate in solid form is the mineral called calcite by geologists, and lime by farmers and gardeners.

A lime kiln, by heating, simply reverses this process to get back to calcium oxide. Calcium oxide made out of calcite and packed into bags is called quicklime. Tip it into the unmarked grave of a murder victim to dissolve away the body parts and bones. Or, even more usefully, add water to make hydrated lime, used in traditional lime mortars. This lime mortar sets by reacting with carbon dioxide from the air to get back to calcite. The toughest mortar has three parts sand to one lime, so that the calcite completely fills all the gaps between the grains of the sand.

Portland cement is heated with added clay, allowing quicker setting by a different chemical mechanism, one that works under water. It is not made at Portland or from Portland stone, but looks a bit like Portland Stone when set. That's not surprising, as Portland Stone is just another sort of limestone, composed of calcite and sand.

Oddly, if water with dissolved calcite gets hotter, the amount of calcite it can contain gets less. This is why calcite precipitates out as limescale on the element of your kettle if you live in a chalky, 'hard' water area. If water containing calcite evaporates, the calcite gets more concentrated and eventually must, again, precipitate out as a solid. A drop of water dangling from a cave roof is evaporating as it dangles, and so may deposit its calcite grain to start making a stalactite.

Seawater washing against a limestone coast dissolves calcite. If such lime-rich water moves into the tropics, and warms up, it can no longer hold all of that calcite. Limestone cliffs here in the UK tend to dissolve into the sea, but in the tropics, the calcite tends to precipitate back out again. The white beaches of the Caribbean are not of normal quartz sand, but of calcite. Each sand grain is a tiny sphere of limescale formed around a speck of shell or grit.

One speck of limescale sand is called an oolith, with both o's pronounced (as they are in 'zoology'). The name means egg-stone, but evokes fish eggs rather than ones from a hen. Compress ooliths into a rock, and it's oolitic limestone, made up of tiny spherical specks. The old, beautiful bit of Bath is built of Jurassic-age oolitic limestone.

Now, ooliths only occur because of that odd property of calcite: as the water gets hotter, the dissolved calcite has to come out of it. So the oolitic limestone shows, as clearly as anything can, that in Jurassic times the UK had tropical, Caribbean-style bathing beaches.

Even in our current non-tropical climate, if lime-rich seawater splashes against the shore, it can evaporate and leave that calcite behind. On the shore of Gower, ancient raised beaches have been glued together with calcite, and now stand high above the present-day beach, cemented onto the Gower rocks. When lime-rich sea water penetrates cracks in those rocks, it refills them with pure white calcite. Calcite usually undissolves into bulgy white lumps. But in such rock cavities, it can be deposited a molecule at a time, forming regular calcite crystals. Such crystal 'spar' was collected in Victorian times and sold to tourists.

The other glue mineral, quartz, is also white, and can also occupy cracks in the rocks to make white veins, but the way it does so is different. The quartz chemical, silica, dissolves in very hot water deep underground. It then squeezes upwards, forcing the rock apart to occupy the gaps. The result looks quite like a crack sea-filled with white calcite.

One way to tell the difference is if there are crystals. Quartz forms eight-sided rods with octagonal ends, while calcite crystals are flat. There's an easier way, but one that has a price on it. That price is a mere 1p, and what else can you get for 1p these days? Calcite is a soft mineral: it comes in at 3 on the Mohs hardness scale, which makes it softer than money, which has hardness 3.5. Scrape calcite with a brown coin, and the calcite comes off on the coin edge. Whereas, if you scrape quartz (Mohs hardness = 7), then a bit of the coin may come off on the quartz.

LEFT Crystalline calcite vein in Mountain Limestone, Worm's Head, Gower

MIDDLE A use for that tiresome brown money clogging up your change. The 1p coin reveals the calcite nature of this vein at Worm's Head.

RIGHT Raised beach from just after the Ice Age, cemented together with seawashed calcite. Gower peninsula.

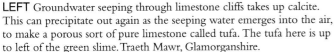

LEFT Groundwater seeping through limestone cliffs takes up calcite. This can precipitate out again as the seeping water emerges into the air, to make a porous sort of pure limestone called tufa. The tufa here is up to left of the green slime. Traeth Mawr, Glamorganshire.

RIGHT Traces of iron dissolved in the calcite give pink and yellow colours in this lump being used as a sea defence at Sidmouth, Devon. The cliff behind is Triassic red mudstone, its particles held together, not too securely, with iron oxide.

BELOW Take calcite, and replace exactly half the calcium atoms with magnesium. The result is a similar mineral called dolomite: a limestone-like rock based on dolomite rather than calcite is also called 'dolomite'. The dolomite layer here probably represents a sea-bed that became a landlocked lagoon with super-salty water. St Monans, Fife.

Calcite is the sea mineral. It's formed in sea water from calcium (dissolved into the sea out of rocks) and carbon dioxide (from the air above). It's extracted from the sea to make seashells; it drops to the bottom as fossils. It dissolves back into the sea to make a mineral cement to stick those fossils together. The resulting rock – calcite fossils glued with calcite cement – put together by Life and seawater is limestone.

We find limestone alternating with the calcite-cemented sort of sandstone along the Jurassic coast of Devon and Dorset and along the other Jurassic coast of Yorkshire. And 200 million years deeper down, at the bottom of the Carboniferous rocks of Chapter 11, we'll find limestone in its purest, finest form: the Mountain Limestone of Gower and Glamorgan.

7. FAULTS

On the right, volcanic basalt – the grey deposits are guillemot poo. On the left, Old Red Sandstone. The basalt is younger, and has been faulted downwards against the ORS. As usual, the sea has got in and washed away the actual faultline. The foreground is volcanic vent agglomerate. North from St Abbs harbour, Berwickshire.

7. FAULTS

Back in Chapter Three, as all the black basalt vents emerged so visibly through the white chalk of Antrim, I promised that it would actually be more complicated than that.

At the end of White Park Bay, tiny Portbraddan stands between two cliffs: black volcanic rock on the right, white chalk on the left. It could be the edge of yet another of Antrim's volcanic vents. However, I've arrived from the west along the base of a couple of miles of basalt cliffs, and two miles would be a bit wide for a vent. The line where the black meets up with the white can't actually be seen. There's a little valley at that point: it runs down to form the harbour. The rock junction is somewhere in the hollow, under all the earth and vegetation.

When we can, we all like to cover up our faults. A fault is where two rock masses have slid past each other. Along the sliding surface, the rock gets broken up – and that's an opening for erosion. So one sign of a rock junction that's a fault is a rock junction you simply can't see.

Still, that inconvenient gap can also happen when the rock contact involves a hot intrusion – like the Bendoo Plug, half a mile to the east and described back in Chapter 3. The chalk next to the arriving hot basalt gets baked and brittle;

the joining-up point, which ought to show what's been going on here, has eroded away.

The third type of rock junction is an unconformity. The older white chalk might have eroded into a natural crag line, then a flood of basalt lava could have pooled up against it. But the basalt doesn't look like lava, as there are no red-topped lava flows (see Chapter 3). And if it were an unconformity, then the junction would probably be plain to see and not eroded away into this little harbour.

ABOVE Portbraddan, Antrim coast: white cliffs on one side, black ones on the other.

RIGHT Looking back between sea stacks and across White Park Bay to Portbraddan. The basalt stack on the right is Elephant Rock (there's another Elephant Rock, made of chalk, 15 miles further west).

BELOW Chalk and basalt at White Park Bay. A major fault has brought basalt downwards on the seaward side. The Bendoo Plug confuses the issue: it is a volcanic vent.

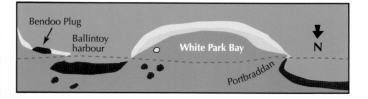

So I walk along White Park Bay, now below cliffs of chalk, thinking it was probably a fault junction but sorry not to have seen the join. At the end of the bay, I round a final chalk cliff, and find a sea full of sea stacks.

The first stack is a rounded arch that reminds me of a teapot – a teapot of white china. But the stack behind is as black as an elephant. From further around the bay, I can look back between the stacks. Every stack to the right of my sight-line is black and basalt. Each one to the left is white chalk. And following that line across the bay, it arrives at Portbraddan, with the white cliff to its left, the black to its right.

The faultline follows the coast east-west; it would be more accurate to say, the coast follows the faultline. But at either end of White Park Bay the coastline wanders north, into ground that's sunk to bring the black basalt down to the sea.

East from the bay, the path stays close to the shoreline, weaving among stranded sea stacks of a raised beach. Then it reaches Ballintoy Harbour. The small harbour faces straight out east. The rocks on its inland side are chalk and those on the ocean side are basalt. The harbour inlet runs straight along the faultline. On the inland, chalky-side, jetty, chalk rock shattered by the faultline movement is just above the harbour stonework.

It's all there, as plain as black and white. Even if it is disappointing not to have laid eye on the actual faultline.

On the coastline of southern Scotland, the same thing happens again. Look across Port o' Warren Bay, and the rocks on the other side are a different way up and a different sort of rock. In the background, the tough greywacke sandstone of the Southern Uplands, its beds raised almost upright by the Scotland-England collision. In the foreground, another sandstone, but a softer one, with water-rounded corners and honeycomb weathering. And this time, the angle's almost flat.

Again, the three possible answers. An unconformity, an intrusion or a fault.

An unconformity is impossible. There's no way the softer sandstone can lie across the eroded-off ends of the greywacke. The angles just won't work. An intrusion is even more impossible: both of the two rocks have sedimentary beds, neither of them has ever been red-hot. So a fault? Walk around the next corner, and you'll not be terribly surprised. Where the faultline junction would be is an eroded-out valley. There's a flat bit at the bottom where somebody's built a shoreline cottage.

Again, the faultline – now I'm pretty sure it is a faultline – more or less follows the shore. Above the shoreline mud, the greywacke sandstone has a somewhat shattered look. Or am

I just seeing what I want to see? Thin seams of quartz ramble through it. A cairn commemorates a long-ago shipwreck, and in the hollow below it, yes, the greywacke is shattered, and also is stained unlikely green with copper minerals. Quartz veins, and unlikely green minerals: both could have been caused by the nearby hunk of Criffel granite, arriving red-hot from underneath. On the other hand, both are symptoms of the friction heat as two rock masses grind along a faultline.

> The western tide crept up along
>   the sand,
> And o'er and o'er the sand,
> And round and round the sand,
> As far as eye could see.

Charles Kingsley's poem *The Sands of Dee* describes the Solway sea creeping back across the mudflats, soothing the silty muck into nice level layers. Somewhere out there a surge of black mud is burying a few mussel shells or a dead flounder: fossils of the future. A rainshower moves on, to reveal a line of white matchsticks against the grey. It's the Robin Rigg windfarm, £325 million, 180MW and 60 turbines: briefly the world's third biggest – one nearly 25 times the size is proposed for

ABOVE Shattered chalk at Port Ballintoy

BELOW Port Ballintoy

RIGHT Port o' Warren, north Solway coast. Near-level brown sandstone underfoot: almost upright greywacke sandstone in the distance. Something wrong here – could it be a fault?

BOTTOM RIGHT Quartz runs through beach pebbles of brown sandstone, greywacke, and also the pink rhyolite that complicates the situation hereabouts.

the Irish Sea. The towers are 80 m high if you're E.on (their owner) or 120 m high if you're a windfarm protester – the owners only measure to the top of the tower. The windfarm winks in and out among the shower-clouds.

Beyond, a grey-black shadow line is the mountains of the Lake District. With the hunched grey rocks below, and almond-scented gorse flowers all along the top, this part of the Solway is a fine clifftop walk even without the excitement of tracking the mountain building moments of 400 million years ago.

Two tall passages lead out either side of a triangular sea stack. You must walk down one of them, or swim down surging grey-green sea water between the high rock walls, before you can see the open sea and the sky. The big coastal faultline slash will also involve smaller fracture movements alongside. Down in the sheltered hollow, where the waves can't bash the evidence, there's a crack in the rocks. The surface to left is smoothed off; a chunk of continental rock sliding past can have that effect. The surface to the right has been shattered, then re-cemented with melted quartz released by the friction heat. It's a 'fault breccia'. Breccia is rock concrete made

with sharp-cornered broken lumps, as against the much commoner conglomerate, or 'puddingstone', where the lumps are rounded and waterworn. Breccia is Italian for breach in the city wall, but can also mean an intellectual breakthrough – as when you work out that this smashed-up rock suggests two bits of a broken continent sliding by.

What sort of crunch-up has been going on here? As King Henry didn't quite say, once more into the breccia, dear friends. It's time for the geological equivalent of measuring the skid-marks left on the road. The next piece of rock is pink: it's a rhyolite vein, squeezed out by the nearby Criffel granite lump. But the squeezing happened before the fault-line movements, because here's a smoothed-off surface, the fault plane itself, slicing across the pink rock. And on that surface, the actual scratch-marks of the faultline movement. It's called 'slickensides' – and that's not Italian or German but good lead-mining dialect, from the Pennines.

LEFT, TOP LEFT TO BOTTOM RIGHT Shattered greywacke with green copper staining; Faultline crack, Gutcher's Isle; Fault plane cuts pinkish rhyolite, Castle Point Bay; 'Slickensides' are faultline scratch marks.

ABOVE Solway coast, from Castle Point. The pink slickensided fault plane of the previous picture is seen left of centre. The cliffs are (mostly) tough old greywacke. The gently-sloping Carboniferous sandstone on the shore is tipped more steeply, the closer it is to the faultline and the greywacke cliff. That's another sign that it's been moving downwards, rather than upwards, against the fault.

The scratch-marks seen here run roughly up-and-down. Alternatively, could they actually be down-and-up? At the end of the bay, there's an abrupt headland called Castle Point. It's an Iron-Age settlement, standing on what I reckon as a small dolerite intrusion. More certainly, it's a fine viewpoint out across the Solway and back along the coast. Children make sandcastles and shriek in the sea, gulls echo the merry noises overhead, and a can of beer just opened hisses on the viewpoint indicator's low table platform. It's just the place to consider dreadful disasters of long ago.

The rock on the seaward side of the faultline is softer. We've walked over it, so we know it is. And its comparative softness is why the Solway is a sea, and Castle Point is dry land and a nice firm stand for the beer can.

Just because the southward side is lower down today, that doesn't necessarily mean that it's the southward side that's slid down. Whether down from above or up from the depths, the southward side is softer rock: that's what's let the sea in. There are two possibilities:

Complex faults at Trebarwith Strand, Cornwall. Right of centre, a reverse fault brings pale Devonian slates over greenish Tintagel volcanics, which are Carboniferous. On the left, a near-horizontal thrust has created 'boudins', sausage-like rolls of rock.

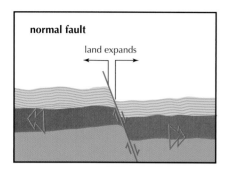

normal fault — land expands

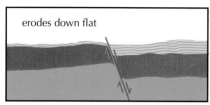

erodes down flat

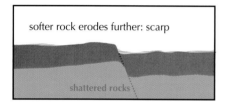

softer rock erodes further: scarp

shattered rocks

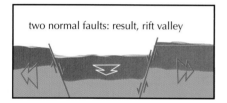

two normal faults: result, rift valley

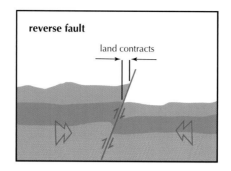

reverse fault — land contracts

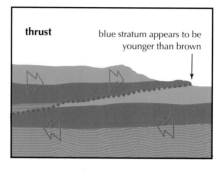

thrust — blue stratum appears to be younger than brown

RIGHT Thrust fault at Three Cliffs Bay, Gower Peninsula

- The soft, brownish rocks on the seaward side started off above the greywacke, and have slid downwards. This makes the brownish rocks *younger* than the greywacke.
- The brownish rocks started off below the greywacke, and have been shoved up against it. There was originally some more of the greywacke on top of the brown, but it got eroded off. Then the sea could get at the brown, softer stuff and wear it away. This makes the greywacke *younger* than the soft, brown rocks. It also implies that, on the landward side, there must still be some more soft brown sandstone buried deep underground.

The crucial question is then: is the brown soft rock younger, or older, than the tough greywacke on the landward side?

The proper way to find out is as follows. Search in the greywacke until you find fossil graptolites. Then fossick on across the yellow sandstone until you find it fossil-free – well, there are some unconvincing black streaks that might be bits of tree. However, along the shoreline the brown rock is interbedded with limestones which yield some smeary looking seashells and at last a definite chunk of coral. For these operations you'll need a hammer, some goggles, a good fossil book, and considerable patience (especially for the graptolites).

The easier way is by experience and feel. If you've knocked about in southern Scotland you will probably know the greywacke. It's the tough ocean-bottom rock crushed by the collision with England in the Silurian period. If you've knocked about in Dorset or Yorkshire, you may get a feeling for the brown sandstone. It's not vivid desert sandstone, it's sludgy and organic. It has a sort of Carboniferous feel to it.

The easiest way of all is the geological map. On the land side: Silurian greywacke. On the Solway side: Carboniferous limestone and sand. The Solway side is about 200 million years younger. It belongs on top of the greywacke, and it's the side that's sunk.

Stretches of time that are also times of stretch

Because I was lucky enough to find an actual fault plane, I know that the brown sandstone slid down off the tough grey rocks in what's called a 'normal fault'. If the fault plane tilts the other way, it's a 'reverse fault'. It makes a difference. Where a reverse fault has been happening, the land has got very slightly shorter. Reverse faults happen when the scenery is being squeezed.

Squeezed scenery gets folded as well as faulted. On the

Solway coast, the greywacke rocks are folded and also faulted – it's suffered the Scotland-England collision. But the brown Carboniferous sandstone is lying almost level and is not folded.

This implies two things:

- As the brown rocks aren't folded and tipped, and the greywacke ones are both folded and tipped, that is another reason to think the brown sandstones are younger than the greywacke (because the folding and faulting happened in between).

- Because one side of the fault (and, indeed, the softer side) is not folded, this North Solway Fault was not a compression fault and is nothing to do with the Scotland-England collision. It's a normal fault.

As land lumps slide apart down the normal fault, the land gets slightly longer. Normal faults take place during times of stretch. The fact that they're called 'normal' suggests that stretch happens a lot of the time. Whenever it's not colliding with somewhere else, a continent is being dragged around the world towards some distant subduction zone. Normal faults allow the continent to respond by getting slightly larger.

Where two normal faults run parallel to one another, the land between them sinks. Even as the land sinks, erosion levels things off again. But it may happen that the fault movement brings hard rocks on one side alongside soft ones on the other. As often as not, old rocks are pretty tough; that's how they got to be so old. The rocks on top are younger and thus, quite often, the softer ones. In that case, the sunken land erodes lower, and it becomes a rift valley.

The UK has so many remnant rift valleys that it isn't worth trying to draw them onto a diagram. North-and-south stretching in the Carboniferous period has given us the Lowland Valley of Scotland, and also the Northumberland Trough and its westward extension, the Solway Firth. The Bristol Channel is another sunken block. East-west stretching in the Triassic gave the north-south river valleys of the Southern Uplands, and the Vale of Eden.

In Chapter 4, sills and dykes were presented as the same items, simply turned around a bit. Sills were horizontal, squeezing into the gaps between the strata – though if the strata had then got tilted, the sills would end up tilted too. Dykes were upright, like stone walls in Scotland.

That's the difference; sills are flat, dykes are up on edge. But the way they get there is different. To slip between the layers, sills must only lift up what's on top. The pressures inside a volcano are enough to do that. But dykes have to actually crack apart the land. That's only possible if the land was planning to come apart in the first place.

Fifty million years ago, the Atlantic Ocean was just being born. England and Scotland were being pulled eastwards, away from the new mid-Atlantic ridge, towards a distant subduction zone. But rocks are solid: however hard you pull, they won't stretch. They'd like to crack apart under the strain. However, the pressures only 5 km under the ground are already 2 or 3 tonnes per square centimetre. That means that gaps simply can't open up. If, though, there's some molten basalt, at 3 tonnes per square inch pressure, ready to rush in and fill it up, then the gap can open.

On the map on page 79, it looks as if one volcano on Mull managed to squeeze out rock as far as North Yorkshire. That's misleading. North Yorkshire, and Cumbria, and most of Scotland, was simply itching to split apart in a north-south direction. The Mull volcano supplied the basalt to fill the gaps.

And so, anywhere on the east coast north of the Solway, anywhere in the west from Yorkshire, you come across a black dolerite dyke. It doesn't make sense in terms of the layers you're looking at, or any nearby volcano. But then you realise it's running northwards: and you go: 'Oh yes. It's a Tertiary dyke, lava from the Mull volcano.'

ABOVE Sea caves and arches are carved along the line of weakness provided by faultline shattering. Study the sea arches in this book (or in the real cliffs) and see, as often as not, the faultline running down through the rockface above. Several rock arches are named as the Needle's Eye; this one is a few miles west of Port o'Warren.

RIGHT Basalt dyke at Sawney Bean's Cave, Ayrshire. A compass can help: this dyke runs just west of north, towards distant Isle of Mull, and is indeed one of the Tertiary dykes from the Mull volcano.

Stones that stop

It can be hard to say what rock it is you're sitting on – is it a rather sandy limestone, or a rather limey sandstone? Is it a lava-flow, or is it a sill squeezed out underground? At the same time it can be absolutely obvious that whatever it is, it's quite, quite different from the rock around the corner. If you're lucky, you can walk up to where the two rocks join together, and stand with your two feet in two different geological eras 100 million years apart.

Edges are interesting: Hadrian's Wall is where Rome ran up against the Barbarians. Lyme Regis is where the eighteenth century meets today's beach tourism. At many places in this book, rocks meet other rocks. This section is a summary.

Conformable junctions
What's happened? Nothing much. A volcano erupts, or stops erupting: a sea gets shallower, and drops sand instead of mud. One rock was being formed: then a new sort starts being formed on top.
Time gap: None.
What rocks: The older rock, underneath, will be sedimentary, or else a lava flow. Rock on top the same.
How to tell: The beds above and the beds below are in the same direction, with no obvious disturbance. The older types of beds and the younger may be interleaved across the join.
Examples: Antrim: basalt lava over chalk (Chapter 3). Sandstone and limestone layers of the Lias (page 138–9). Marloes Bay: Old Red Sandstone over Silurian underwater sediments (page 194).

Unconformities
What's happened? A new rock has been laid across the eroded-away surface of an older one.
Time gap: The rock underneath is older: possibly much, much older.
What rocks: Rock underneath, any sort; rock on top, sedimentary or lava flow.
How to tell: The junction will be bumpy, with the newer rock trickling down into gaps in the older one. Often the older rock was eroded off by an invading sea, which then laid down the younger sediments: the base of the younger rocks may contain beach pebbles made of the older ones. If the older rock has been tilted or folded before the arrival of the younger, then the old beds are in a different direction: a nice, clear 'angular unconformity'.
Examples: Hutton's Unconformity, Berwickshire: Old Red Sandstone over Silurian greywacke (pages 11, 13). Jurassic Coast: chalk and Greensand over various Jurassic, and New Red Sandstone (pages 127, 140). Glamorgan: Jurassic limestone over Carboniferous limestone (page 187). Portishead: New Red Sandstone over Old Red Sandstone (page 171).

TOP Grey underwater sediments lying on Old Red Sandstone, east of St Bride's Haven. The junction is conformable – the sea level rose and so a new sort of stone formed.

BOTTOM The Glamorgan unconformity forms the cave roof: Jurassic white limestone above Carboniferous Mountain Limestone at Dunraven Bay

Dykes, sills, volcanic vents
What's happened? Volcanic rocks squeezed into surrounding countryside.
Time gap: The intrusive rock is younger than the country rocks below *and* above.
What rocks: The intrusive rock will be igneous, with crystal structure usually too small to see. The country rocks can be any kind.
How to tell: Dykes and sills often show columnar jointing, upright in sills, but sideways in dykes. Where sills intrude into sedimentary rocks, they slip between two beds, with smooth top and bottom surfaces. The edge of the intrusion cools quickly, forming a glassy sort of stone, often with a slight colour difference. The nearest few centimetres of the country rock may be baked into brittle, unstructured 'hornfels'.
Examples: Antrim: basalt vents (Chapter 3). Various dykes and sills (Chapter 4). Mull volcano dykes (previous page).

other side of the junction, the country rock is baked to 'hornfels', structureless, brittle, and sometimes with new crystals of its own. The result can be that, close to the junction, the intrusion and the country rock come to look more and more like each other. There may well be quartz veins (and copper and lead mines) above the intrusion.

Examples: Land's End granite (page 80). Gabbro intrusions, Pembrokeshire (page 71, 214).

Faults

What's happened? Two rock masses have slid past each other, bringing rocks of different kinds up against each other.

Time gap: The downfaulted rocks will be younger; but it's usually hard to work out which side is downfaulted.

What rocks: Can be any kind. If you've already worked out what the rocks are, this may reveal which is younger and what the faultline movements have been.

How to tell: Rocks along the fault plane are shattered and cemented back together again, forming fault breccia. The shattering may mean that the faultline itself is an eroded-out gully, with the junction not visible. Look out for smaller, parallel faultlines alongside: and tangles of quartz veining released by friction heat.

Examples: White Park Bay, Antrim (this chapter). North Solway Fault (this chapter). Marloes Bay (page 211).

Large intrusions

What's happened? An underground lump of molten rock, many kilometres across and many kilometres deep.

Time gap: The intrusive rock is younger than the country rocks above and alongside.

What rocks: The intrusive rock will be igneous, with visible crystals: typically granite or gabbro. The country rocks can be any kind.

How to tell: Spot the crystal structure of the 'plutonic' intrusion. Close to the junction, the intrusive rock cools more quickly, and so the crystals are small or even invisible. On the

ABOVE Dolerite sill, Balcomie Links, Fife. The poorly defined columnar jointing in the dolerite is at right angles to the bedding of the sandstone below. If you climb up there, the top surface of the sill is notably flat.

RIGHT Dolerite dyke at Bearraig Bay, Skye. The edges cooled more quickly and are slightly orange.

TOP, FAR RIGHT Granite edge in sea defence boulder, Aberdeen. The molten granite has squeezed a vein into the country rock, a grey schist.

BOTTOM FAR RIGHT Fault breccia along the great Solway Fault

8. JURASSIC COAST

Arratt's Hill, just east of Beer. In the distance, the Great Unconformity brings the chalk directly down onto red Triassic mudstone. On the headland itself, the Seaton Hole fault brings the chalk right down to sea level. What's missing from this picture of the Jurassic Coast? The entire Jurassic period.

8. JURASSIC COAST

The Jurassic Coast of Dorset and Devon is confused. It starts firmly in the Cretaceous or chalky period: perhaps it's been watching Steven Spielberg's 1993 film about a family being chased by *Velociraptor mongolensis* – the film really ought to be called *Cretaceous* not *Jurassic Park*.

Spielberg should also have made the Velociraptors roughly knee-height, and covered in feathers. Authentic as it might be to be pursued by an intelligent turkey, the Jurassic Coast starts in the east with the magnificent (if Cretaceously chalky) sea stacks of Old Harry and his Wife.

Once past Swanage, though, the coast is limestone. And limestone all the colours of an English afternoon tea: honey-coloured, and cream-coloured, and the colour of the freshly baked farmhouse loaf. Across the clifftops the grass is limestone grass, short and lawnlike, speckled with wildflowers. But at the lawn edge, where there ought to be a herbaceous border, there's empty air, and somewhere down there a sea that's clear green, floored with white limestone sand. On the shoreline shelves, the giant ammonites are golden, and the fossil trees of Lulworth, outlined in what was originally a green algal slime, are transformed to limestone in teatime yellow.

That golden colour stretches inland. It surfaces as the stonework of Somerset's market towns, the sunshine glow of Salisbury Cathedral, the honey-coloured oolite of Bath and a thousand Cotswold cottages. This is the colour of the Jurassic, and it stretches from the Dorset coast north-eastwards all the way to Yorkshire. The Jurassic is the golden heart of England; and when you come across Jurassic anywhere else – the Glamorgan coast of Wales in its enclave below the coalfields, the Jurassic beaches of the Isle of Skye below black basalt ridges – it's a shock and a misplacement.

William Smith: main man of the Jurassic

Today, we know the good way to keep track of the different rocks. Take an ordinary flat map, and paint over it in different colours. Red for the Old Red Sandstone, pale green for the chalk, scarlet for granite. That's the way it's been for 200 years, and brief as that may be in geological terms, it's still long enough for us brisk humans to get totally used to the idea.

But in 1799, no such map had ever been drawn – scarcely so much as thought about. Then into the mind of one man, William Smith the canal surveyor, there sneaked the

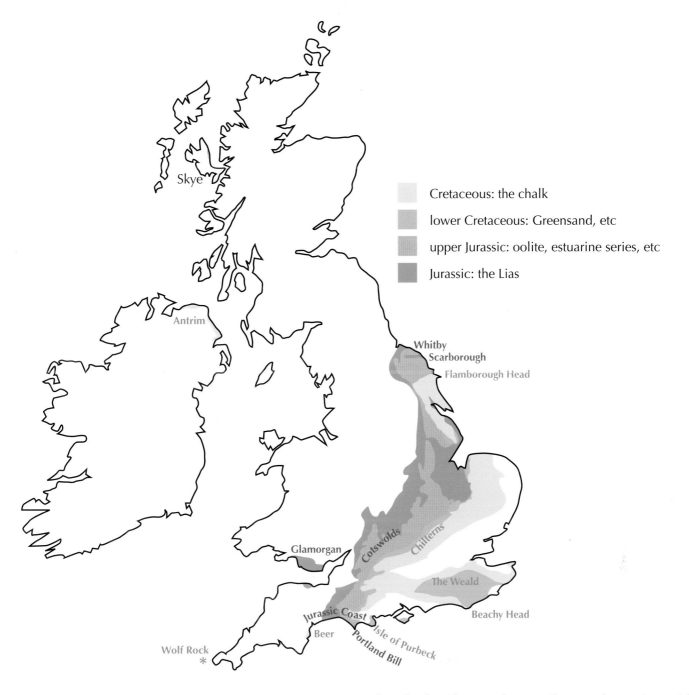

Cretaceous: the chalk

lower Cretaceous: Greensand, etc

upper Jurassic: oolite, estuarine series, etc

Jurassic: the Lias

Skye

Antrim

Whitby
Scarborough
Flamborough Head

Cotswolds

Chilterns

Glamorgan

The Weald

Beachy Head

Jurassic Coast

Beer

Isle of Purbeck

Portland Bill

Wolf Rock
*

LEFT Old Harry Rocks at the eastern end of the Jurassic Coast

ABOVE Having traced the chalk from Old Harry Rocks and Brighton, through the Chiltern Hills to north Norfolk, the Wolds of Lincolnshire and Yorkshire, and Flamborough Head; the Jurassic from Dorset by the Cotswolds to the Cleveland coast of Yorkshire: there does seem to be some sort of pattern.

thought that if one took a set of watercolour paints, with a sufficient range of different and distinctive colours and tones, and marked on the various and different rock formations so painstakingly observed in the canal banks: the Inferior Oolite, the Cornbrash and the rest – that then perhaps, just perhaps, the result might be in some way intelligible, what one could without exaggeration describe as a geological charting, or map.

It's easy to mock the overheated style of Simon Winchester's *The Map that Changed the World*. With hindsight, the idea of a geological map is pretty obvious. In the eighteenth century, the idea of a geological map was untested, unthought of – but still pretty obvious. Doing it was the difficult bit.

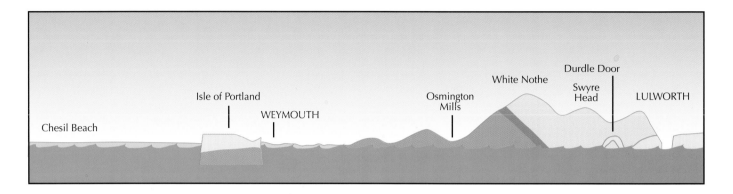

To make it even harder, William Smith was not a country parson with a private income and plenty of country parson friends he'd met at Oxford. William Smith grew up on his uncle's farm – a farm that just happened to be in the Cotswolds, and that, accordingly, just happened to have fossil sea urchins used as doorstops and bryozoan seashells rolling around the fields. Young William pestered his uncle for books, educated himself in mathematics and natural history, and aged 18, got himself apprenticed to a surveyor. In 1791, he was surveying at Stowey under the Quantocks – in the village five years later inhabited by Coleridge and Wordsworth. He missed the two poets, but did spot the same red marl as in Worcestershire, with the same Lias rocks lying below. And he must have made a good job of the survey. His next job was on the estate owner's mines in the Somerset coalfield, between Bristol and Bath.

A mine is a challenge for any surveyor, let alone one on his second independent commission. It's not just the lack of fields of vision and sightlines, but also the intricacies of working in three dimensions. For Smith, it was a look at Somerset from underneath. In the chained basket he dropped all the way down what he would later call the 'stratigraphic column'.

Somerset's coal rush required its own canal to compete with canal-carried coal of the Midlands. The surveyor's job went to the self-possessed, rather quiet, Mr Smith – who perhaps also charged a bit less than his rivals.

In 1799 (just as Napoleon seized power in France) Smith got hold of a map of a 10-mile circle around Bath. Its white spaces cried out to be coloured in geologically. And as he rode (his landlord charged him half a crown a week extra for his horse) across Somerset, he examined his canal and peered over riverbanks. He mentally joined up sandstone and limestone through the hills and under the fields, until he'd painted in every rock found over those 30 square miles. Just 16 years later (as Napoleon was dealt with at Waterloo), he'd done the same for the whole of England and Wales.

The difference between sandstone and limestone is obvious enough. Except when the limestone goes from creamy to yellowish, from smooth to textured, and transmogrifies itself from a sandy limestone to a limey sand. Tracing rock strata under woods and rolling pasture to the

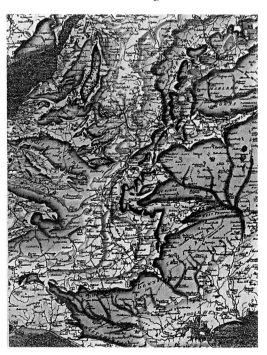

FAR LEFT Smith's Stones: a stratigraphic column erected at Coombe Hay on the Somerset Coal Canal. The rocks are selected from William Smith's 'Order of Strata' compiled at Bath in 1799. From bottom to top: Carboniferous Pennant Stone; Triassic White Lias (2 blocks); Jurassic Blue Lias (greyish, 2 blocks), Inferior Oolite, Great Oolite, and Forest Marble (2 blocks); Cretaceous chalk. To stop the whole thing from toppling over, 17 other strata identified by Smith have been omitted.

LEFT A detail from Smith's map of 1715 shows the chalk (green), Jurassic limestone (yellow) and the Lias (blue) setting off northeastwards from the Dorset Coast. Poole Harbour is at bottom right.

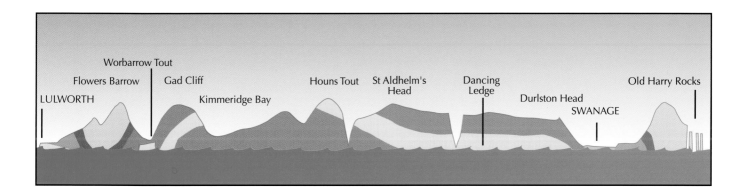

quarry half a mile away is straightforward enough, when the layers lie level. But they don't; they tilt down into the ground. Worse, they bend. Worse still, they come to a fault-line and stop, and start again 100 ft underground.

The Principles of Steno, listed in the first chapter, were: superposition, what's underneath is older; original horizontality, what slants was flat to start with; lateral continuity, what's over here is also over there. These were crucial, but not enough. What made Smith's task possible were those sea-urchin doorstops, and little pebbly shells. It worked because of the fossils.

Smith's Law of Faunal Succession

1. Fossils are not curiosities, or chance marks in the rocks. They are the remains of living beings.
2. Each creature lives just once. If it stops in the strata (becomes extinct) it doesn't restart.
3. If two different rocks – a sandstone, say and a limestone – contain the same fossil creature, then they are the same age.
4. If two rocks look the same but contain different sets of fossils, then they are different rocks.

Smith is dubbed the 'Father of English Geology' for his one-man mapping, over all England and Wales, and also for his ground-breaking (strictly, rock-breaking) realisation about the fossils. But for the second of these, some credit goes also to Jurassic England.

Along the Dorset coast, eastwards from St Aldhelm's Head, each bay competes with the others to show you something more surprising. Durlston Bay has its fossil mammals, small shrews lurking in chalk holes waiting for something horrid to happen to the dinosaurs. The small creatures did well to cower. Keat's Quarry, inland from Dancing Ledge, has dinosaur footprints over a metre across. To the west of St Aldhelm's, Kimmeridge Bay reveals ammonites and reptiles, on a dangerous tidal shoreline. Here too are fossil bacteria in the liquid state known as petroleum; Kimmeridge has the UK's only onshore oil well, and oil also seeps into the sea. And just east of Lulworth, there's a fossil forest: not the trees themselves but rings of algal slime that built up around their bases, now solidified into limestone.

Jurassic Coast: key

- sand and shingle
- Cretaceous: chalk
- Cretaceous: Greensand and Gault
- lower Cretaceous: clay and sand
- Portland Stone (within upper Jurassic)
- Mid and upper Jurassic: sandstone, shale
- Jurassic: Black Ven Marls
- Jurassic: Blue Lias
- Triassic: red beds
- ●●● Budleigh Salterton Pebble Bed

As you explore this coastline, one particular tough, pallid limestone becomes a well-recognised friend. It's the Portland Stone, and it's been lifted on edge by the Alpine mountain-building to form every scenic feature between Peveril Point and Portland itself. At the eastern end of Worbarrow Bay it runs out to sea as Worbarrow Tout. It emerges again at the bay's other end as the Mupe Rocks. At Lulworth, the sea has broken through the tough, upright bed. Once behind, it has hollowed out the perfect scallop-shell shape of Lulworth Cove.

At Stair Hole, just to the west, the Portland Stone appears in an intimate, dishevelled state. Again here the sea has broken through behind, to show the tangle of rock delight-fully named the Lulworth Crumple. Another mile westwards, and the same bed makes Dorset's finest sea feature, the flying buttress of Durdle Door. The Portland bed diminishes into a series of stacks: the Bull, the Blind Cow, the Calf. But it'll be back in six more miles, bigger than ever …

The Isle of Portland is a peninsula (this still makes it more island-like than the Isle of Purbeck, which has sea to the south

Life and Jurassic times of the Portland Limestone

ABOVE Mupe Bay: easily-eroded chalk forms the bay

BOTTOM Mupe Rocks, at the west corner of Mupe Bay, looking across to Worbarrow Tout. Both are of the hard white Portland Limestone.

BELOW Tough Portland Stone, raised on edge by the distant arrival of Africa, forms the sea arch at Durdle Door.

BOTTOM Lulworth Cove. The sea, having broken through the tough Portland Stone, forms a near-circular bay in the softer chalk behind.

of it only). Portland, projecting from the exact centre of the World Heritage Coast, is the ideal seaside spot to take a break away from the stress and business of your everyday life. Well, it is if your everyday business is knife fights and stealing cars: Portland's almost-island is the site of Verne Prison, and also of the Portland Young Offenders' Institution, its harsh grey masonry of 1848 matching the earlier castle of Henry VIII.

The peninsula, along with its buildings, is the definitive arrival of the tough limestone layer that's been nudging the coastline since Swanage. Portland limestone is an oolite: seen close up, its texture resembles fish roe, or thousands of tiny eggs. Each egglet has at its centre a particle of shell or sand. As the sun heated the subtropical Jurassic sea, calcite precipitated around the central fragment just as limescale precipitates in your kettle when you heat the water. Gentle wave action formed the particles into egg-shapes, before they glued together into a solid mass. The resulting stone is 'freestone': it can be shaped into blocks or carvings. It's tough enough to resist weathering, but not so tough as to be unworkable. Sir Christopher Wren chose well when he built St Paul's Cathedral out of it, so the architect of HMYOI Portland was only following a well-established trend.

Cheaper cuts from the quarry are weakened by the casts of former seashells, where the actual fossils dissolved away to leave spaces in the stone. This defective stone can be seen in sea defences and wall copings all along the Dorset coast.

West from Portland runs the Dorset coast's truly unique feature. Chesil Beach is 14 miles of flint pebbles. This makes more flint pebbles than most of us require, so I hurry past Chesil, to the place where early nineteenth-century fossil hunters and early twenty-first-century beach babes form a single happy tradition.

Lyme Regis

> The young people were all wild to see Lyme… As there is nothing to admire in the buildings themselves, so the remarkable situation of the town, the principal street almost hurrying into the water, the walk to the Cobb, skirting round the pleasant little bay, which in the season is animated with bathing machines and company, the Cobb itself, its old wonders and new improvements, with the very beautiful line of cliffs stretching out to the east of the town, are what the stranger's eye will seek; and a very strange stranger it must be, who does not see charms in the immediate environs of Lyme, to make him wish to know it better.
>
> Jane Austen, *Persuasion*, written in 1815 and set in 1814

The two hundred years since *Persuasion* have added charm also to the buildings Austen didn't think much of. Add to that the ammonite streetlights – and also the Cobb's literary appeal, important in two separate classic novels: the crucial opening moment of John Fowles' *The French Lieutenant's Woman*, and Louisa's near-fatal leap from the steps in *Persuasion* itself. The Cobb walk, its ancient cream-coloured stonework, its graceful curve out from the front of the town – it's as irresistible as ever. But looking down at the twirly gastropod shapes etched into every inch, you have to ask: how come observant Jane Austen missed the fossils?

For in 1814, as Louisa made her fateful leap from the stone steps, Lyme was already a world centre for the fossil fashion. Five years earlier, a lady walking on the beach had paid a little girl called Mary half a crown (about £10 today) for an especially fine ammonite. And two years after that, that same child had uncovered the world's first plesiosaur. Anne and Captain Wentworth arrived in the off-season: else the talk of the town must surely have been the monsters of the rocks being uncovered by Mary Anning.

A Portland screw is the local name for these handsome gastropods.

(for key see page 131)

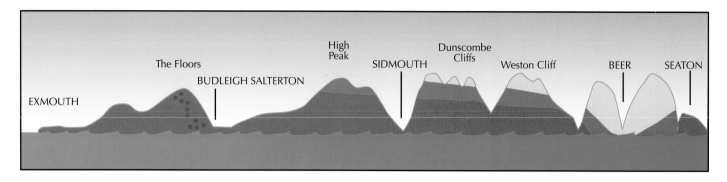

ABOVE Lyme Regis, with its ammonite lamp standards

'She sells seashells by the seashore'

Mary Anning was a pioneer palaeontologist – or at least she should have been. She discovered the first plesiosaur (or underwater dinosaur) and the first ichthyosaur (dolphin-like, but with a vertical tail instead of a flat one). But she was a woman, and a working-class woman at that. Accordingly, her discoveries were credited to the academic gentlemen of Oxford and Cambridge who wrote them up and stole them from her.

That, at any rate, is the twenty-first century story. Anning wasn't a palaeontologist, a scientist knowledgeable in past life. Anning was a professional fossil-hunter, and a skilful one. The ichthyosaur went to a local collector for a mere £23, but in 1830 she sold the new plesiosaur for 200 guineas. (An ichthyosaur today costs about the same in real money: £13,000 including free delivery.)

The social conventions of the early nineteenth century made it difficult for a gentleman and a working person even to hold a meaningful conversation. But the parsons and academics who studied the rocks often put their enthusiasm above the social conventions. William Smith had difficulty pursuing the strata without a private income, but he did eventually succeed. A generation later his nephew John Phillips moved smoothly into a professorship at Oxford.

Some of Mary Anning's clients treated her as a mere hireling servant, which was common at that time – Darwin once threw his hired beetle-collector down the stairs for selling to a rival. But to several she was a valued colleague. Henry de la Beche was a gent: his father owned a slave plantation in Jamaica. But he made friends with Anning when they were teenagers in Lyme. It was while explaining to Mary the scientific side of her many discoveries that de la Beche painted a fanciful monsterscape of Jurassic Dorset. When Anning fell on hard times, he had the painting printed off as a lithograph, and arranged for her to receive the proceeds.

We do have to blame the general social opinion of the age for the way that, while a lower-class fellow like William Smith just could be, women were not accepted as academic geologists. Or we would have to, were it not for Etheldred Benett in Wiltshire. This expert in fossil sponges and molluscs donated a coral fossil to William Smith, and was praised in public by Roderick Impey Murchison, president of the British Geological Society. The Emperor of Russia, assuming 'Etheldred' must be a man, conferred on her an Honorary Doctorate of Civil Law at St Petersburg University, where women were not even admitted as students, let alone as Honorary Doctors.

Fossil hunting is as fascinating as ever it was. Hire a proper geological hammer – the stylish ones are painted yellow. Wander along the beach until you find a stone – there are plenty of stones. Hit the stone with the hammer. After five minutes, you've knocked a few chips off the stone. After ten minutes, you've knocked a few chips off your fingers. But with any luck you haven't put your eye out – although Adam Sedgwick did, not wearing goggles as he hammered at Robin Hood's Bay in Yorkshire.

Some of those rounded limestone pebbles contain beautiful fossils. But most of them don't. The best way to see the ones that do is at the Charmouth Heritage Coast Centre, which offers advice, hints, and a duty warden for any interesting and inscrutable finds. Downstairs, if you're desperate, there's a commercial fossil shop.

But the best way to see fossils for yourself is to walk along the beach, slowly, without a hammer in your hand. Look at the seawashed boulders; fossils there are too well anchored to be carried away by other fossil hunters. Eye up any freshly broken bits of the black shales fallen from Black Ven behind the beach. Small, portable fossils will only be found if you're first on the beach as the tide goes down, or better still, after a

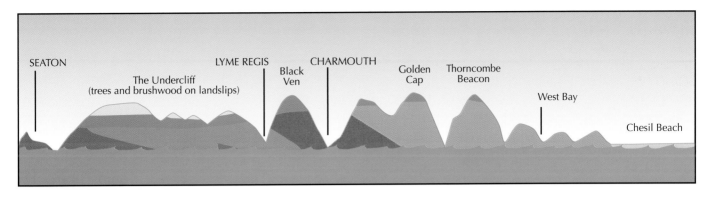

SEATON   The Undercliff   LYME REGIS   Black   CHARMOUTH   Golden   Thorncombe   West Bay   Chesil Beach
(trees and brushwood on landslips)   Ven   Cap   Beacon

ABOVE The Cobb, Lyme Regis

BELOW Fossils in Portland Stone of the Cobb, Lyme Regis. A trace fossil in the cementwork provides the clue as to why they don't get a mention in Jane Austen's *Persuasion*. Moulded rubber boot soles date the stonework renewal to no earlier than the 1950s.

winter storm. But at any time at all, cover a mile of the foreshore, at a gentle pace, and you are guaranteed to see some ammonites.

Sit down on your beach towel, and finger-sift a patch of beach gravel. Crinoid segments, and small belemnites, are the likeliest finds. And if you have to hammer, look for the shaly pieces that fall apart if tapped gently along their edge. And don't forget, when you hire that hammer, hire some goggles too.

The Layers of the Lias

We hove our ship to with the wind from sou'west, boys
We hove our ship to, deep soundings to take;
'Twas forty-five fathoms, with a white sandy bottom,
So we squared our main yard and up channel did make.
Sea-chantey, *Spanish Ladies*

The 'white sandy bottom' isn't a prophetic reference to the nudist beaches at Budleigh Salterton, Durdle Door and Studland. Mariners in days of sail attached a lump of wax to the lead used for depth soundings, so as to examine a sample of the sea bed. Their charts marked the sand, gravel and so on.

Even without the chart, the kind of bottom could still give an indication of the distance from the shore. Close to the land there'll be coarse sand and gravel, even quite big pebbles. Beyond that, only the smaller sand particles; and beyond them, fine silt drifts to the sea floor. Even further out, where no land sediments can reach, there may be shells of free-swimming sea creatures, and a thin ooze of white lime.

If we're looking at ancient sea bed that's now up in the open air, we're looking at a sea bed that, at some point, has been rising. The upper strata will have formed in a shallower sea. At the bottom, clear white limestone with seashells in it. Above that, fine mud turned to mudstone, siltstone, and shale. Above that again, tough sandstone, perhaps with pebbles.

But why stop at sea level? An estuary has been bringing those pebbles and sand. As the shingle builds up, it forms a seaside swamp of tree-ferns. Seen from 400 million years afterwards, that shows as coal. Or there might be a sea-edge lagoon, whose stagnant, oxygen-starved waters will give a sort of sandstone that's greyish-black.

Westwards along the Jurassic Coast is, in rock terms, downwards. West of Lyme Regis we reach the bottom of the Jurassic, the rock layers called the Lias. Lias even means 'layers', and the Lias shows alternations of shale and limestone better than any rock has the right to.

Ask what's going on here, and I would explain about shallow sea giving mud, and eventually shale; deeper, clearer water leaving just the fossil shells of its sealife, and the pure calcite dissolved out of them – which is, eventually, limestone.

'But how come all those grey and yellow layers?' So I explain again, more slowly, trying to pretend it was my over-complicated explanation rather than just you being stupid. But you rather rudely interrupt. Yes, you understand all that. But the Lias of Glamorgan, seen on the following page, has 130 layers of limestone and sandstone. Was the sea going up and down like a kid on a pogo stick?

It seems that it may have been; so long as you think of a very slow pogo stick. During World War One, the Serbian Engineer Milutin Milancovitch was interned by the Austrian

army. Instead of cursing his fate, he spent the time working out three different wobbles in the orbit of the Earth. His calculations were ignored until 1924. Then they were taken up by two broadminded weather forecasters; one of them being Alfred Wegener.

The longest and most important Milancovitch cycle is the 100,000 year one. The earth's orbit is an ellipse. But due to gravitational nudges from Jupiter and Saturn, that ellipse varies between being almost circular, and rather more oval. The total sun heat reaching the earth over a full year does not vary. But when the orbit is at its most oval, the arriving heat is concentrated during four months of every year, and reduced during the rest of the time.

The main astronomical climate effect is called winter, and it's caused by whether the North Pole is pointing away from the Sun. The ovalness effect is smaller, and at the moment, it works to counteract the winter-summer effect (in the northern hemisphere, that is). The earth is slightly further from the sun during January, and slightly closer in during July. This means that here in the northern hemisphere winters are milder than they might be, and our summers slightly chilly. Meanwhile in Australia and Antarctica, the

LEFT Duria Antiquior, 'Very ancient Dorset'. The species so vigorously eating one another 200 million years ago were all later to be uncovered by Mary Anning. Note the ammonites, belemnites and crinoids, and even the plesiosaur poo, later to take fossil form as a coprolite.

summer is slightly more summery and the winters slightly worse. (Slightly better, of course, if you're a ski-er in the Snowy Mountains.)

The difference in arriving sunshine due to the ovalness of the orbit is only 6 per cent. That's because we are currently at an almost-circular stage of the cycle. In 50,000 years, with the orbit at its most oval, there will be, over the course of each year, a 25 per cent difference in arriving sunlight.

So what if England's winters are three per cent warmer, balanced by our summers being three per cent cooler, while Australia works the other way around? It turns out that our slightly warmer winter, with slightly cooler summer, encourages snowfall in the spring. Snow is white, and reflects incoming sunlight straight back into space. (To melt the snow on your vegetable patch, spread soot over it.) So slightly more snow makes the ground slightly cooler again, resulting in slightly more snow again. It's a vicious snowball. When there is land close to the Pole ready to receive this snow (as there is at the moment), small climate wobbles get amplified, and an ice age can start up suddenly.

The shale-limestone alternations in the Lias do seem to tie in with the 100,000 year Milancovitch Cycle. During the Jurassic, there was no land close to either of the poles; also, the Jurassic was a warm time. Textbooks tell us there were no ice ages during the Jurassic. However, recent climate modelling using computers suggests that there may have been small Jurassic icecaps on the southern continents, waxing and waning with the 100,000 year Milancovitch Cycle. Such icecaps would extract water from the oceans, and lower the mean sea level by a few metres. (Floating ice, on the Jurassic Arctic sea, would not have any effect on sea level. An iceberg displaces exactly its own weight of seawater.)

ABOVE Black Ven, between Lyme and Charmouth

BELOW Ammonite graveyard, Black Ven

Milancovitch cycles cause small ice ages in the Southern Hemisphere. The ice ages lower the ocean. The computer also thinks there's more monsoon rainfall in the northern tropics at this time. The salt water overlying England responds by switching from a clear blue limestone sea to a muddy shale one. This is the currently least unconvincing theory of the Lias layers. That very slight wander in the Earth's orbit has made clear seas alternate with muddy ones, as regularly as the pendulum of a great-great-great grandfather clock, over the 130 cycles and 13 million years it took to form the cliff in the picture.

This Lias is in South Wales. The Blue Lias is at Lyme Regis, and we'll see Lias again at Staithes in Yorkshire. The Lias does seem to be a 'universal formation', a rock layer over, or under, all England. Then again, this does look quite like the alternating shale/limestone from the Carboniferous period seen in Northumberland on page 186. Given time and the perceptive eye of William Smith, we could explore across England, finding the connecting Lias outcrops all the way between. We could puzzle out the chalk that lies on top, and the New Red Sandstone underneath (but itself on top of that confusing Carboniferous).

The easier way has also been given us by Mr Smith. The Devil's Toenail fossils of Glamorgan immediately say 'not Carboniferous'. The same Satanic feet-snippings appear at Staithes in Yorkshire. The Toenails are not liars and the corresponding rocks are, accordingly, Lias.

LEFT More than 130 layers of the Lias at Traeth Mawr, Glamorgan

TOP Blue Lias at Devonshire Head, Lyme Regis

MIDDLE Lias at East Quantoxhead, Somerset

ABOVE Devil's Toenails (the bivalve *Gryphaea*) distinguish the lower layers of the Lias, whether in Yorkshire or Dorset or here, at Traeth Mawr. These are in the grey shale layer at eye level in the left-hand picture. One wouldn't want to stand around under the 129 unstable layers above poking out the bivalves.

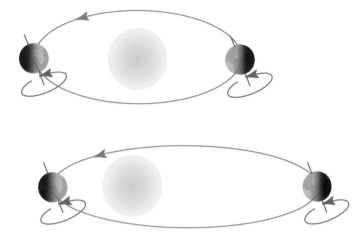

The Great Unconformity

The sandstone and the greywacke at Siccar Point; the pale grey Carboniferous limestone and the pale yellow Lias limestone, at Southerndown. Every such unconformity confronts the shortlived viewer with the abyss of time: hundreds of millions of years for tilting it over, and eroding away the ends, and a whole new layer of rock being laid in a whole new ocean. The biggest unconformity in all England runs along the cliffs of Dorset and into Devon, 35 miles long from Abbotsbury around to Sidmouth. It's spotted again at White Nothe above Kimmeridge Bay; inland, it runs from back across Hampshire as far away as Eastbourne. It tracks the invasion of the chalk sea described in Chapter 6. It's called, appropriately, the Great Unconformity.

From the start of the Jurassic coast at Old Harry, we've been passing gradually downwards in time, lower and lower in the Jurassic. Which is another way of saying, the strata have been tilted upwards to the west. But suddenly, those slightly slanted strata run only half way up the cliff. Above that, their slightly tilted ends have been sliced away, and it starts all over again with the Cretaceous. And this time, the Cretaceous hasn't been crumpled up by the Alpine mountain building. This time, the Cretaceous lies conveniently flat.

Immediately above the unconformity are two non-chalky layers – non-chalky, but genuinely Cretaceous, as they lie

ABOVE The Milancovitch 100,000 year cycle.
The two Earths on the left are experiencing January. The northern end of Earth's axis is tipped away from the sun, so the northern hemisphere gets extra darkness and fewer sunbeams. Right hand side: July, and it's winter in Australia. In the bottom half of the diagram, Earth's orbit has become oval. This adds some heat in January (England's winter, Australia's summer) and cools down July (England's summer, Australia's winter).

BELOW The Great Unconformity: chalk, upper Greensand (yellow, but growing trees) above Mercia Mudstone: Haven Cliff, Seaton

TOP The distinctive Pebble Bed pebbles can be carried by longshore drift as far east as Hastings in Sussex. The red colour is superficial. A broken one shows the tough white quartzite, washed out by flash floods from a mountain range somewhere to the south that is now, apart from these pebbles, completely vanished.

BOTTOM Ammonite from the Blue Lias at Devonshire Head, Lyme Regis

smooth and straight against the chalk beds above. The lower layer is the Gault: a soft stratum of clay. Above the Gault is the Greensand. It's a sandstone, but it's only green when you break it open: the surface is a lively yellow colour. Both Gault and Greensand are soft rocks, content to crumble. Most of the way along, the Greensand is authentically green, being overgrown with vigorous trees. The trees mean the unconformity isn't so striking close up. It's when you look along the coast from many miles away that you see the strange inconsistency in the cliffs.

Just before Beer, there's a jump in the rocks. It's a major fault, though the line of it is, as usual for faults, hidden within an eroded-out bosky vale called Seaton Hole. And from there for a while westwards, the cliffs are completely chalk. But after Beer, everything starts to turn red.

The so-called Jurassic Coast, having flirted full length with cretaceousness from Old Harry Rocks right along to Beer, veers off upwards and breaks off at clifftop level. While up out of the sea to displace it, there rise the older rocks of the well-named Red Coast. These may still be part of the Jurassic Coast World Heritage Site, but they are Triassic rocks.

The walk west along the Dorset coast has been a walk downwards through the rocks, deeper and deeper into time. Old Harry was near the top of the Cretaceous, a mere 65 million years ago; the dinosaurs who wandered the nearby landmass were unaware of how shortly they were to be displaced so humiliatingly by the sort of furry small life in burrows they trod on without noticing. At Portland, the yellow-brown cliffs were early Jurassic, 140 years old. The dinosaurs who left their backbones buried in the mud were at the height of their dominance – and any thinking dinosaur who predicted dinosaurdom would be top of the world for

ever, was absolutely correct. (If we primates are still on top in 100 million years, we'll be doing *very* well.)

Seatown is not to be confused with Seaton, 10 miles to the west and on rocks 10 or 20 million years older. West of Seaton, red mudstone from 200 million years ago rises at the back of the beach. Above it, the Great Unconformity still recolours the upper cliff. Seaton cliff is the slices of a Neapolitan Ice Cream: red base, yellow (but tree-covered) Greensand, white chalk. The difference is, the ice cream machine turns out its slices straight, while the Triassic mudstone is at an angle. And the time gap at the unconformity is now a full geological period; the entire Jurassic has existed and been eroded away again between (as it were) the raspberry flavour and the banana.

Budleigh Salterton

The conventions of beach nudity are simple. It's rude to peer. It's better to be a couple than a solitary male; more tactful to take your own clothes off; and the camera should stay in the rucksack.

Does the local council take evil glee in siting the nude beach signs to make you nervous? At Southerndown, the naked bathers were placed underneath a spectacular cliff collapse of the Glamorgan Lias. At Budleigh's naked place, the pebbles are potato-sized, and awkwardly steep: uncomfortable

to lie down on, and worse to walk down in bare feet (not to mention bare everything else) to the sea.

The nude-beach convention about cameras makes it embarrassing to photograph the Pebble Beds, even if those rounded stones from 300 million years ago do attractively match the rounded bodyparts of the beach people beneath. The quartzite Pebble Bed pebbles are oxide-stained in the warm tones of one who forgot the suntan lotion. Having peered, non-intrusively, over one's fellow bathers, it's even more intriguing to look down at the sea-brightened pebbles. They were carried by flash floods out of a mountain range, somewhere to the south, where today there is only the sea. By 1794, the Somerset Coal Canal had been surveyed from end to end. In the brief break before construction started, William Smith was sent north to examine the other canals. It was a splendid chance to check his Somerset strata outwards into the rest of England. The stagecoach took him up the outcrop of the Jurassic rocks, to York. He climbed the tower of York Minster, and spotted Dorset chalk re-emerging along the Yorkshire Wolds.

It was this that encouraged him to go public with his theories, communicating them, rather nervously, to the amateur naturalists of Bath. In social terms, this was a climb across the unconformity between the dark industrial Coal Measures and the gently sandy strata of the Jurassic. At the start of the nineteenth century, the gap between a gentleman and anybody else was almost as stark as that between people and animals, between a bird and a tree.

William Smith, with his salary of a guinea a day, had clambered part-way across the financial gap. Any of Jane Austen's heroines would have rejected an income of just under £400 a year, but it was reckoned a gentleman with no dependants and frugal habits could live on that amount – about £26,000 in today's terms. Smith got a middle-class mortgage, and bought himself a three-storey house at Tucking Mill, southeast of Bath.

The study of natural history offered a persistent social wriggler a wormhole into the gentlemanly realm. The country parsons and minor aristocrats who studied plants, insects and rocks were already the oddballs of their class. Many – possibly even most – were prepared to speak to the alarmingly knowledgeable Miss Anning, the earnest and intriguing Mr Smith, as if to one of themselves – at least so long as the topic was rocks and fossils.

Smith would eventually make it into the world of gentlemanly geologists. But it would take him most of his life. That life journey would, it turned out, be a rerun of that coach journey of 1794. It would take him up the fossil-bearing Jurassic strata, the Lias and the Oolite, north and slightly east across England; from the sea cliffs and ammonites of Dorset, right up to the ammonites and sea cliffs of the Yorkshire coast.

Budleigh Salterton Pebble Beds

142

9. YORKSHIRE ROCK

Long Nab: gently rolling North Yorks Moors end abruptly at Cleveland's steep sea cliffs.

9. YORKSHIRE ROCK

There is, perhaps, no place in the world where ... a geologist may revel with more pleasure and satisfaction, than along the Yorkshire coast ... Although Granite, and other unstratified rocks, exist in our district only as boulders, the basaltic dyke, which may be well examined a few miles from Whitby, gives no bad idea of rocks formed by fire ... The lofty hills towering above the spectator with their frowning summits, as in the neighbourhood of Whitby and Scarboro', but more especially at Peak and Boulby, or corroded into fantastic forms of ruined castles and magnificent columns as at Flamborough head, and the deep and solemn caves of the Lias, hollowed out beneath by the billowing waves of the ocean, leave on the mind the sublime sentiments of vastness, grandeur and awe.

The Fossils of the Yorkshire Lias, 1855;
Martin Simpson, Whitby Museum curator
(as quoted by his successor Roger Osborne in *The Floating Egg: Episodes in the Making of Geology*, 1999)

As William Smith moved across Somerset, digging coal mines or canals, the same rock layers kept reappearing, in the same order. Those layers dipped gently to the east. A day's ride towards London, strata that had been on the surface were now buried deep, and new strata had appeared out of grey middle-England skies (as it were) to lie on top.

But how far could he go and still find it the same? Smith's first idea was that the layers lay slantwise, one on top of the other, right around the circle of the world – like the ridge-line pattern on an ammonite. The rock layers could have been tipped that way by the rotation of the Earth.

This idea is attractively simple – but it doesn't make sense. As you rode your pony eastwards across Europe and Asia, the rocks would be younger every day; younger again across America; and younger still as you finally arrived at – the very rocks you'd started off on. Mr Smith, who never liked theoretical geology anyway, soon gave up on that one.

Cliffs east of Whitby: Estuarine Series over alum shales

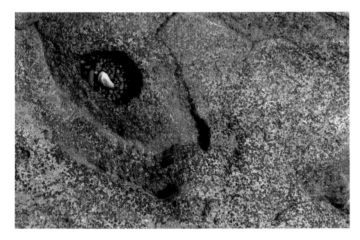

LEFT Ironstone nodules at Hummersea Beach. Lias limestone and shale above.

TOP Ironstone nodule, Crook Ness. The sandstone beds arrange themselves around the ironstone lump, indicating that the concretion formed itself as the surrounding sandstone was hardening into rock.

MIDDLE Shelly fossils in ironstone, Hummersea Beach, are another indication that this mineral ore has never been a red-hot seeping vapour.

BOTTOM Oolitic, meaning egglike, ironstone (Scalby Bay). The millimetre spheres (mini-concretions) resemble fish roe rather than hen's eggs.

Even so, how far *could* you go, outside the edges of Somerset, and still find the same stones? In 1794 he got a chance to find out, when he was sent to study the canals of England before starting the digging on Somerset's own Coal Canal. The answer turned out to be, surprisingly far.

A clear blue sea lays down Oolitic limestone in the Isle of Purbeck. The clear blue Caribbean is doing the same today, across 300 miles of the Bahamas Banks; and it turns out that the clear blue Jurassic sea of England was much the same size, its honey-coloured limestone stretching up across Oxfordshire, though Northamptonshire, to reach the sea again south of Scarborough. The Lias, alternating sand and sludge of a dirtier sort of sea, rises in the cliffs of Devonshire Head, at Lyme Regis. It rises again at Whitby; it breaks off and falls onto the sunbathers exactly the same in Yorks as in Dorset. 'Lias' is the local quarrymen's dialect for 'layers' – in Yorkshire geology books those are Yorkshire dialect quarrymen, but in Dorset it's supposed to have been the West Country way of saying it …

And travelling inwards towards London, upwards in time, you shook off the coal dust and crossed first the Lias, and then the golden oolite of Bath, and then the Oxford Clay; until you came to the greenish sandstone that Smith named the Greensand. Eastwards, and further up the levels of the rocks, and you came to the Marlborough Downs, and the chalk.

So when, in 1819, Smith drew a section of the strata from London right across to Snowdon, it was, near enough, a section from London across to Somerset, or from London northwards into Northumberland. And it's interesting to head off to Yorkshire, to look at Dorset, as it were, from the other end.

Cleveland Coast

Leaving aside places like the Thames, no coastline has been worked over by Man like the Cleveland coast of the North Yorkshire Moors. There has been fishing ever since the Stone Age, and today there's beach tourism. But in between those times, there's been mining of iron, digging for jet, and the removal of whole clifftops in the quest for aluminium sulphate, or alum. Even today, the place boasts Britain's second deepest hole, the Boulby Potash Mine; and down near the bottom of it astronomers lurk listening for the neutrino whispers of dark matter in the empty spaces of the universe.

But to the Vikings Cleveland was Cliff-land: tall crumbly cliffs of mud and shale crumbling into the sea. They crumble slowly: every dozen years, the Cleveland Way footpath moves another metre in from the edge. But that's still quick enough to make you feel the movement of the sea inland, the erosion away of old rocks and the piling up of a new stratum of sandstone under the sea. Meanwhile, down below the crumbling cliffs, every high tide washes out ammonites and oysters of 200 million years ago, and sometimes even a dinosaur. In the occasional notch, preserved by the sea cliffs that hang over them and wall them in, sit pretty little fishing villages like Staithes and Robin Hood's Bay.

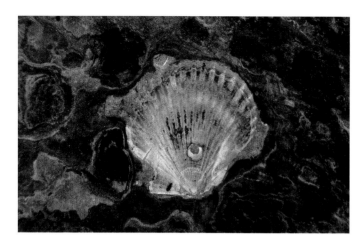

ABOVE Scallop fossil at Hummersea beach

BELOW RIGHT Saltwick Nab: the upper part of the peninsula has been completely removed as an alum quarry.

BELOW LEFT Alum crystals, Whitby Museum

Bus services out of Whitby allow the coast to be divided into convenient day-walk chunks. Everywhere from Saltburn round to Flamborough Head can be covered in a wandering week, interspersed with frequent crab sandwiches. Saltburn is a proper seaside village, with an esplanade and a proper pier. Northwards, though, the pier-stroller gazes towards strange industrial structures of Redcar and Teesside. The water of Skelton Beck is rusty red; and where the stream crosses the shingle, the waves come back in bloodstained brown. The old ironstone mines have been leaking since 1999.

High on Hunt Cliff, an interpretation board marks a Roman signal station that was half there in Victorian times but has now completely fallen into the sea. Unusually for this coastline, Skinningrove is not a self-consciously pretty little fishing village. It's a ironstone mining settlement, built in brick and breeze block in every style from Late Victorian to 1950s.

The iron itself isn't hard to find. Blobs of ironstone stick out of the brown cliffs. Red, rounded knobs of it emerge from the wavecut platform like the skulls of the damned that Dante walked over in the glaciated Ninth Circle of the Inferno.

The ironmasters didn't care how the stuff had got there. They just wanted to get it out again, and send it away to the Midlands to supply the Industrial Revolution. But if we're familiar with metal mining, this ironstone is puzzling. Ores of lead and tin and silver are boiled out of the rocks by nearby lumps of granite magma. They seep upwards through cracks in the rock, and condense again as mineral veins, often intermixed with quartz.

These iron oxide rocks aren't like that at all. They occur in individual lumps, not in uniform veins. Often they contain seashells, which implies they've never been melted. This ironstone formation has happened under the sea, and in two stages.

A rock with a bit of iron in it, such as a red desert sandstone, washes away as sand into the sea. Certain bacteria happen to find iron useful, and absorb it into their little bacterial bodies. As a consequence, rusty reddish sludges form at certain places on the sea bed. And then, as the sand is squeezed and slightly heated towards becoming sandstone, the iron molecules migrate. Iron oxide, like most molecules, is gregarious: it likes nothing better than to grab onto another iron molecule. And so, slowly, iron concentrates itself into these lumps that now sit within the sandstone like rather brownish and flattened cherries within a cherry cake.

In Chapter 6 the same thing happened to silica minerals, snuggling up within the chalk sediments to form flints. Calcite, the limestone mineral, often forms similar lumps

A tower, made of Alum at the Sandsend Works and given as a wedding present in 1862 to Mr & Mrs Thornton Presented to the museum by Miss Thornton 1957

within sandy rocks. The lumps, of whatever mineral they may be, are called 'concretions'. Sometimes they form around an initial lump of something: a salt marsh in the Wash (Lincolnshire) has built ironstone concretions around abandoned World War Two munitions. The silica and calcite ones can form around fossils, and so get enthusiastically bashed apart by geologists.

East of Skinningrove, you can descend an iron ladder to reach Hummersea beach. Or you can walk the cliff foot from Skinningrove, on a wavecut platform of rock slabs with patches of green waterweed. The cliff is crumbly and it's best to walk well out from it. There are ironstone lumps, and many fossil shellfish, including some easy-to-see scallops.

East again, the cliff path rises over Rockhole Hill, nominated by VisitYorkshire as the highest sea cliff in northeast England, and one of the highest in the UK. At its original height of 200 m it may have been. But today it descends in two steps, with a level shelf at two thirds height. That's not natural, but comes from the quarrying away of an entire geological rock-layer: the alum shales.

Alum Money

Filter it gently and put aside the urine with your collection of secrets.

Rhazes, tenth-century Persian physician, philosopher and alchemist (quoted by Roger Osborne in *The Floating Egg*, 1999)

A stream bed running through old alum workings at Sandsend Ness is stained red by the iron still in the quarry tailings.

As you sit on Rockhole Hill enjoying your picnic, your child squashes some raspberries into the nice white tablecloth, and your old father knocks over the wine. You smile and try to make light of it; and the fact is, those stains will clear, just about, with a couple of times through the washing machine. What rinses in, rinses out.

But this doesn't apply to the cheerful colours in your tee-shirt, and it doesn't apply to the stripes in your beach towel. This is because of a particular chemical called a mordant, which attaches chemically to both the dyestuff and the fabric being dyed. From the Middle Ages onward, England's main industry was wool. To make your fortune from wool, you don't export it raw; it should be spun and woven into cloth, and also dyed. This requires a mordant called alum. Europe's only alum works were owned by the Pope. This was troublesome, when Henry VIII was rebranding England as a Protestant country.

Alum Bay, Isle of Wight, has impressive yellow sandstone and London Clay of Tertiary age, as well as chalk, all raised upright by the Alpine mountain-building – but no alum. The name came from the fact that it was the site of early, misguided exploration for alum.

Early in the seventeenth century Sir Thomas Chaloner, traveller and intellectual, went hunting on the Yorkshire moors – and noticed rocks which resembled the Pope's own alum shales. He was granted a Patent, an exclusive licence to mine, by James I. When James' son Charles granted 'exclusive' licenses to some other people as well, Chaloner was so miffed that he turned Parliamentarian and ended up as one of the judges who signed Charles' death warrant. This wasn't the first royal contretemps because of alum. Henry VIII may have married Anne of Cleves for the sake of her access to the Dutch alum mines. But on the wedding night she turned out to be not pretty enough, so Henry annulled the marriage, paid her a big pension, and did without the alum. The term 'alimony' could be (but isn't) derived from this episode.

If you ever need to recreate civilisation from scratch, the single most useful study would be GCSE chemistry. The clifftops quarried down to the sea; the landowners enriched; the port of Whitby made prosperous by boatloads of wood ash and human urine: all because nobody in the seventeeth century had taken GCSE chemistry. Alum is aluminium sulphate. Sulphur is not scarce, and aluminium is either the commonest or the second commonest element in the earth's crust. All you need to make alum is to attack with sulphuric acid any clay mineral (with its aluminium in it). But if you don't know Standard Grade chemistry, you have

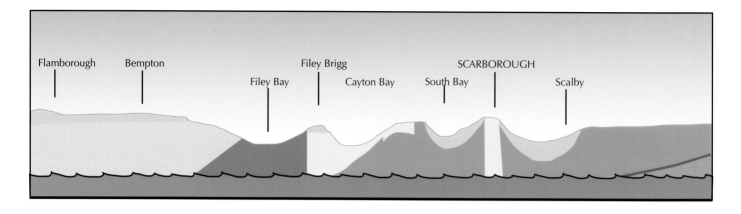

to find an obliging rock-source that performs the operation of its own accord.

Alum shales contain sulphur in the form of pyrites – the iron sulphide mineral (FeS) that creates golden-black fossil ammonites, or the sparkling crystals of fool's gold. The shales contain water of crystallisation. Thirdly, they contain fossil carbon, very weak coal, no good for your fireplace but concentrated enough that once alight, alum shales will smoulder slowly for the necessary couple of months. And, fourthly, they are mostly made of mud and, accordingly, have aluminium.

Set the alum shales on fire, and let them smoulder for nine months. The iron sulphide oxidises (burns, in effect) to become iron sulphate; some of that transforms into sulphuric acid. (Some of it doesn't, leading to difficulties at the next stage.) The sulphuric acid attacks the aluminium clays to make alum.

Dissolve the result in water, thus leaving behind the silicate part of the aluminosilicate clays. And then, extract the alum crystals without at the same time extracting the nasty black iron sulphate. The 'alum-makers' secret' involves a floating egg. And it's explained in lively detail in the book of that name by Roger Osborne.

What about the wood ash and urine? Well, alum isn't actually aluminium sulphate, it's a double sulphate of aluminium plus either ammonia or potassium. So add either urine or ashes for that.

In 1798, a trainee apothecary from Penzance called Humphrey Davy found himself with enough private wealth, and consequently leisure, to amuse himself with all the varied intellectual delights of the early nineteenth century. One of his friends was Mr Wedgwood, the man with the pottery factory. Two other friends, Mr Coleridge and Mr Wordsworth, were busy revolutionising poetry. While they experimented with new verse forms, Davy experimented with gasses, finding out how much carbon monoxide he could breathe without dying. With the two poets he tested the intoxicating effects of laughing gas. He managed to set fire to a diamond, and invented the safety lamp that would save coal-miners' lives over the following century. His scientific fun led him to the discovery of no fewer than six of the chemical elements (sodium, potassium, calcium, magnesium, boron and barium) plus most of Standard Grade chemistry.

Some people find GCSE chemistry pretty tough stuff (but then, some people aren't that interested in geology …) For them, Humphry Davy realised in 1815 that about half of your two-year course could be conveyed in four words, plus three symbols:

$$acid + base \rightarrow salt + water$$

Among much else, this explains alum. Take bauxite, which is aluminium oxide or hydroxide, and so is a base. Pour on sulphuric acid, and get alum. In 1831, a vinegar trader called Peregrine Phillips came up with a cheap way of making sulphuric acid. And as this became an industrial process, the alum mines of Yorkshire, one by one, closed down.

Staithes

The Cleveland sediments arrived after Britain's first two mountain-building shocks. Yorkshire's far enough north that the third one, the Alpine crunch, merely squeezed and stretched it. Accordingly, Cleveland's cliff layers lie more or less level. Along the northern coastline, that's a roughly level layer of tough sandstone, lying on top of roughly level shales and mudstones, alum and ironstone and jet. A tough top, and the sea carving into the softer bottom, leaves Cleveland with cliffs almost vertical, almost uniform all the way around.

Here and there a stream breaks through the tough upper layer. It carves a cliff notch into which is jammed a fishing village. A steep path between the blackthorn bushes widens to become the descending street of Staithes.

In the rising triangular gap, the houses of Staithes lie on top of each other like boulders from a cliff collapse. They are built of the same brown stone as the layered cliffs of Penny Nab and Cowbar Nab above. But stone is simply a brown base for orange roof tiles – clay naturally pigmented with iron oxide for a colour scheme that brought 40 artists to Staithes in the nineteenth century. Did some of that artistry scrape off onto the locals? Traditionally, houses were decorated in the colour schemes of their boats, so that when wreckage was washed ashore, everybody knew whose boat it had been. But residents today have chosen the exact washed-out sky blue that completes the orange and brown colour scheme.

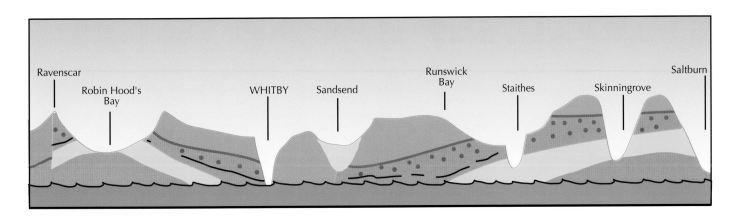

Cleveland Coast: key

	boulder clay
	Cretaceous: chalk
	lower Cretaceous mudstone
	upper Jurassic sandstone, limestone
	Estuarine Series
—	The Dogger ironstone
••	ironstone
•	alum shale
▬	Jet Rock
	Staithes Sandstone
	Redcar Mudstone

Why pantiles? Roofing slate is a metamorphic rock, the compressed result of mountain-building. The UK's slate is quarried from our most ancient rocks in Snowdonia, the Lake District and Glen Coe. Before modern roads, heavy slates could easily travel only by sea; and all of those mountain quarries are on the wrong coast for transport to Staithes.

On the level ground behind the clifftop stands modern Staithes. Arriving by the cliff path, we didn't discover it until the following day; but for motorists it's the real place and old Staithes remains a mere remnant down in its hole. Real Staithes has car parks and buses, street lights and a small supermarket. It's built of brick, and breeze block, and slate, and concrete tiles.

But the streets of old Staithes run down, not up – to the harbour, sheltered between its two high Nabs. In the

Staithes, one of England's prettiest villages. High house prices and low local wages mean that there are now very few permanent inhabitants. Cowbar Nab is river-delta sandstone, scrambled by worms and seashells. The dark igneous rocks along the jetty are erratic boulders, but travelled by boat rather than glacier.

early nineteenth century, with the North Sea still full of herrings, this was the busiest fishing port in northeast England. The handsome stone piers are half-buried under blocks of black rock: a necessary precaution, when a storm two centuries ago took 13 houses, including the one once lived in by the eighteenth-century explorer Captain James Cook. The black storm-barrier rocks are obviously alien. The big crystals in them indicate an igneous rock, and one which cooled slowly, in a magma chamber deep underground. The black colour and the crystals suggest a gabbro – unthinking, I failed to test with my compass for the mineral magnetite.

'Where do they come from, the black barrier rocks?' My landlady thought it was a trick question, I already knew the answer. The UK gabbro is on the Isle of Skye; sea travel from Norway seemed more plausible. She thought they might have come from Norway.

Behind the piers, around the corners of the two Nabs, the cliffs extend for ever. Staithes Sandstone was laid down in a shallow sea. It shows lots of cross-bedding, considered to be caused by storm waves surging back and forth against some nearby sea cliff. But the cross-bedding is itself obscured by lots of bioturbation, caused by the creeping through of shellfish and worms. These rock details aren't hard to find. On the road up out of the village, you press yourself against them whenever a car comes by.

The tough sandstone alternates with softer silt – once again it's the Lias of the last chapter. The sandstone layers hold the cliff together at the top. Down at the bottom, though, the softer siltstone layers let in the sea. So the cliffs are as steep and sheer as anywhere else in Cleveland. The waves creep across the rock platform, to trap any preoccupied geologist against the cliff. As the wave-cut platform carves ever inland, bits of cliff fall down to display their ironstone nodules and their fossils. And on occasion they crash down on someone, like poor Miss Grundy in 1802, whose head was separated from her body by a falling stone.

Back in the village, Staithes Beck wiggles inland between the houses. Even without a view to the harbour, you hear the tide creep into the village at midnight, as the boats lift off the mud and bang together. Our B&B was up the village, in a street just wide enough for two walkers to pass. Outside were the chimney pots of the house below, and a red roof-line with two seagulls who squawked at each other all night. An ammonite was propping open the bedroom window.

Whitby: Fossils, goths, and the world's first jet aircraft

Whitby is a mix of the sinister and the silly. With its ruined abbey against the moon, it's the historic haunt of Dracula. As well as the grim but fictional vampire, it is currently also haunted by Goths in dark flowing dresses and heavy eye makeup, and that's just the chaps. But when it comes to bringing in undead corpses carefully preserved in graveyard earth, Count Dracula has to yield priority to a Jurassic fossil crocodile …

Where the coastline cuts across a dome in the strata, as it does at Robin Hood's Bay, you can see all the various layers in rings, spaced outwards from the centre. And where the coastline cuts across a centuries-old settlement such as Whitby, you get the same effect in human, rather than sedimentary, building materials. At the old river port, the axis of the pericline, there's a conglomerate of fish boxes and orange plastic in a matrix of yellow stone. The charming cottages of the seventeenth-century fishing port no doubt discharged sewage straight into the seawater to leave the narrow streets neat and clean.

Today all's converted to chip shops and amusement arcades, not to mention the Dracula Museum, whose crimson neon twinkles again in the harbour water as the tide creeps up between the streetlights. High above, the moon competes with the gaunt arches of the floodlit abbey. Salt winds have worn down the sandy gravestones to pick out the burrows of fossil shrimps.

A century further out is the stratum of Georgian town houses from the more prosperous eighteenth century, when Whitby was the main port of eastern England. That prosperity was founded on sea-going urine tankers for the alum industry. Well, there was also wood ash. Further out again, later in time, rises a bed of high-fronted hotels; and a strange mineral accretion of beach huts. Beyond those again, the grey concretions of today.

ABOVE Whitby harbour

RIGHT Fossil shrimp burrows exposed by wind erosion in the Abbey graveyard

BELOW Penny-in-the-slot jet workers, Whitby Museum

BOTTOM LEFT Fossil plesiosaur, Whitby Museum

BOTTOM RIGHT The world's first ever jet aircraft, Whitby Museum

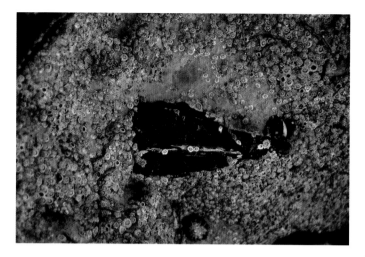

A volcanic intrusion cooks and hardens the surrounding country rocks. The countryside around Whitby has suffered 'contact metamorphism' from the urban intrusion. West of the town, the altered clifftop countryside consists of a golf course, and to the east, a caravan site.

Whitby Museum is a museum of how museums ought to be. Here are glass cases crammed full of stuff, some of it illegibly labelled, some of it not labelled at all. They have a Hand of Glory cut from a hanged thief and pickled black. They have a mannequin headhuntress, fetchingly dressed in skins and a spear, accessorised with a papier-mâché severed head. They have a penny-in-the-slot machine that elucidates the manufacture of dreary jetwork jewellery.

The jet itself is fossil monkey-puzzle tree, floating out to sea and then, when waterlogged, sinking into mud where it could be preserved. This makes it a form of coal. It's easy to carve and to polish, and light enough that you can wear an awful lot of it at once. This makes it ideal ornament for mourning wear; in addition to Victoria's Prince Albert, the nineteenth century suffered high child mortality.

Old jet mines can be seen as small holes along the shore, some of them below the high-tide line. Just because it was such a well-known jet site, Whitby is now entirely mined out. What you see in Whitby's jet-propelled trinket shops is all imported, and for this book's jet picture I went to the Jurassic coast of south Wales.

Whitby's alum quarriers and jet miners sometimes found strange lumpy bits in the rocks. In 1758 came the first incident of people (Capt. Chapman and Mr Wooler) finding these things interesting and writing to the Royal Society. By the 1820s, the fossil monster industry was well established at Lyme Regis, and a fossil crocodile from the alum shales was sold to the Whitby Philosophical Society for £7 – about £600 in today's terms. But the market really hotted up by 1841, when a bidding war broke out between the Whitby Philosophers and Cambridge Museum. A 17 ft ichthyosaurus eventually went for £230. Today, a 15 cm fragment of Yorkshire ichthyosaur could be yours for £550. But behind the cabinet with the world's first 'jet' aircraft, Whitby Museum displays a world-class collection of Jurassic fossils.

Walking southwards around the Cleveland coast is walking very slowly upwards in time. Above the Lias, according to William Smith and his Somerset strata, we should find the Inferior and the Upper Oolite, the shallow-water limestone that built Bath. But we find nothing of the sort. The clifftops are a coarse, cross-bedded sandstone, interbedded with black shales. It's named as the Estuarine Series; but the estuary in question was actually a delta. Braided delta channels deposit banks of gravel and sand – hence the cross-bedding. A sub-stream blocks itself up with its own gravel banks, and moves into a new channel; and the former channel fills up with swampy organic sludge. Those are the dark shales.

Careful tracing of the sandbanks indicates that the river was flowing south-westwards: out of the North Sea and into Yorkshire. And this makes sense of the Dorset disparity. With the North Sea a raised upland, and the sandy rivers flowing down into the sea at Whitby, Dorset would be 100 miles or more offshore, in clear unsedimented seas. We needn't be upset that what's above the Lias and underneath the chalk isn't Oolite. Once we've hunted out some plant fossils, and found them the right Oolite age, then any residual distress is eroded out.

My first Yorkshire stopoff was at Saltwick Bay, immediately east of Whitby and just below the caravan site. Within 20 minutes I came across a black, broken ammonite. It was dawn, before any other ammonite fanciers; and the tide was just starting to fall – I had the first look after two separate high tides had turned over the shingle. On the other hand, I wasn't seeking ammonites, just watching my feet over the wet boulders while carrying camera and tripod for an early-morning photo.

While waiting for a cloud to pass off the sun, I also found the belemnite cast shown on page 41, plus some other ammonites; and a lesson in the instability of the Lias. Warnings about cliffs, tides and slippery boulders are part of today's overcautious age. Proper geologists ignore that sort of stuff, surely? Well, they do, and Adam Sedgwick lost an eye while chipping out a fossil; William Smith paralysed his legs by falling off Castle Rock at Scarborough, and Mary Anning lost her pet dog Tray under a cliff landslide. And as I waited for the sunshine, a tipper-truck's worth of shale dropped quite suddenly to the shoreline, just as the geology book said it would.

The cliff structure east of Whitby is two-tier: tough Estuarine Series along the top; softer Lias layers below; and hordes of kittiwakes perched along the join. Tough top and erosion underneath makes for a nice steep cliff, which is the safe sort if you're a kittiwake, and the stripy Lias offers ledges for nesting.

Five miles from Whitby, Wainwright's Coast to Coast path arrives at the end of its cross-country hike from St Bees on the Permian red sandstone coast of Cumbria. The path continues above the high cliff of Far Jetticks, where the wind raises dust and grit off the bare ground of the alum quarries. Around the bend of North Cheek there are views across Robin Hood's Bay with its village slotted into the back corner. The two Cheek cliffs demonstrate how an anticline, an upwards fold in the rocks, can shape a bay. Understand these strata, and you'll have to conclude that once, high above Whitby Harbour, modern streets with supermarkets were eroded away by wind, rain and the river. The bay itself will delay you with either its fossils or else its ice cream van.

My great-great grandmother came from Scalby Ness, before she married Sir John Johnstone's handsome (but slightly serious) land steward. However, I was after stuff a whole lot older than great-great granny. A walker on the morning shoreline called her friend to look at a large rock. The friend wouldn't come back over the slippery boulders. But I was walking that way anyway.

I thought it was a piece of fossilised firewood: a scrap of tree from 200,000,000 BC. That's what she thought it was as well. 'And you know, I did once find a dinosaur footprint, right here on this beach.'

BELOW LEFT Cross bedding in the Estuarine Series, east of Whitby. The angled cross-bedding indicates strongly moving water: the dark-coloured strata are from stagnant swamp with rotting organic matter. The alternation of sand and sludge indicates the shifting outflow streams of a great river delta.

BELOW RIGHT Robin Hood's Bay shows how an anticline, an upwards fold in the rocks, can shape a bay.

The books say 'Dinosaur footprints can be found' – but I've always taken that to mean 'but not by you, mate'. It's much, much more convenient to spot them on the warm, dry (but rather poorly lit) rockfaces of Whitby Museum. But this woman had found one within half an hour of coming down the steps. Mind you, that was seven years ago. And she'd not found another since. Within minutes of her finding it, a collector had spotted the way she was excitedly studying it and hauled it away. 'I was really hoping he would drop it on his toes. But he didn't.'

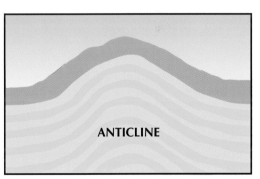

ANTICLINE

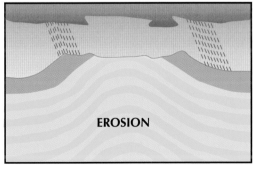

EROSION

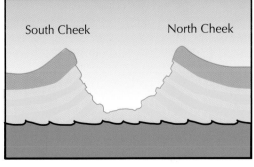

South Cheek · North Cheek

LEFT Fossil timber, Crook Ness

RIGHT Dinosaur footprint, Whitby Museum

FAR RIGHT Scalby Ness to Scarborough. One of the best bays in England for dinosaur footprints, but you'll have to be sharp to get ahead of the footprint snatchers.

Scarborough

William Smith could have enjoyed a brilliant career as a civil engineer and drains man. 'This was a tide in his affairs, which, had he followed the middle current without stopping to examine the banks, would have led him on to fortune,' as his nephew John Phillips wrote. Unfortunately, the banks of his life were Jurassic sandstone, and had fossils in them …

In 1799, William Smith lost his guinea-a-day job in Somerset. Had they noticed that he was spending his survey time peering into canal cuttings and quarries? From now on, he would be a freelance. That was good in terms of surveying the strata at the same time as surveying the actual surveying. But in terms of making a living, it was a disaster. Even though his known expertise let him raise his daily rate to two guineas and then to three, his travelling expenses (up to 10,000 miles a year) meant he was unable to keep up the mortgage on his slightly gentlemanly Somerset house so full of rocks and fossils.

In 1815, the end of Napoleon's wars brought a fiscal crisis. The national debt demanded serious financial cutbacks, as well as the (purely short-term, of course) monstrous new charge, the income tax. Two of the items wealthy patrons found they could manage without were grand ground improvements and the associated surveying, as well as advance subscriptions to Mr Smith's interesting (but expensive) map of rocks. Meanwhile his wife was suffering mental health problems, or what the nineteenth century more simply described as going mad. Despite this, and perhaps remembering his own upbringing, he took on the care of his orphaned 15-year-old nephew John Phillips.

Nephew John turned out to be not a nuisance at all. There are teenage boys who collect stuff such as fossils, and love the tops of stagecoaches and downmarket inns where one might even meet a highwayman. Young John would be Smith's geological sidekick for twenty years of wandering across the rocks of England, and eventually became Professor of Geology at Oxford and President of the Geological Society.

Smith's great geological map of England and Wales eased his difficulties only briefly. Within months, pirate editions were on the streets. He was forced to sell his 3,000 fossils, amassed over 20 years, to the British Museum. This put the museum in an embarrassing position. They were not used to geologists who actually needed the money. But Smith found an ally in one gentlemanly geologist, Mr William Lowndes, who happened also to be Chief Commissioner of the Tax Office. The aristocratic Sir Joseph Banks, pioneering botanist and first director of Kew Gardens, was another friend and supporter.

The £500 for his fossil collection and the occasional advance of £50 from Sir Joseph Banks was not enough. In 1819 his creditors had him committed to debtors' prison in Southwark. Smith served his time at the Kings Bench Prison – where another inmate was Mr Micawber, the character in Charles Dickens' *David Copperfield*, who spends a couple of pages of Chapter XI behind its spiked wall. Debtors' prison was strangely sociable and open; after a morning in tears at the shame of it, Micawber spent the afternoon in a game of skittles. Smith was imprisoned for 11 weeks, before the sale of his house paid off the mortgage and he was released into the street. Homeless as he was, he was still highly employable as a drainage expert.

For seven years he was a wanderer, picking up surveying and drainage work along with plenty of fossils. He reinforced the sea defences of Norfolk using natural-style banks of sand and pebbles. When the hot springs of Bath ran dry, Smith plugged the underground leak. His nephew and sick wife travelled with him. In 1820, they stopped at the health spa of

Scarborough in hopes of helping his wife's condition (plus, of course, the chance of spotting ammonites and dinosaur footprints). 'He imbibed for it a partiality which augmented with future knowledge,' as John Phillips was to put it in his biography of his uncle.

In 1822 he was denied, for the second time, membership of the Geological Society. But in 1824 he was back in Scarborough to give a public lecture ('Geology a Science of great extent and universal interest; not a science of hard names, but of beautifully according facts.—The great facilities for acquiring it afforded in our own country.')

'Always charmed with the bold and varied line of rocky coast, and interested by the geological peculiarities of the north-eastern part of Yorkshire, he gladly seized on this occasion of renewing his residence,' was how John Phillips put it. But Smith, now 55 years old, must have been pleased with the people as well. The provincial gentlemen of the Scarborough Philosophical Society were eager to catch up with Oxford and Cambridge, and thrilled to have Mr Smith in their midst.

The Philosophers of Whitby had opened their museum two years before, complete with fossil crocodile. But the Whitby Museum was, to put it frankly, a jumble (and still is). Scarborough was to have the world's first purpose-built

In 2008 the Rotunda was reopened with the original Smith scheme restored in its circular gallery. The directors didn't on that occasion consume their museum in the shape of cake. But the philosophical ideal continues. You can bring in a strange stone and they'll do their best to tell you what it is – even if, as often as not, it's a tooth from one of the beach-ride donkeys. Even the tradition of feeling slightly superior to Whitby is maintained: they aren't terribly impressed by *The Floating Egg*, written by Whitby's volunteer curator, and don't stock it in their shop.

Meanwhile Sir John Johnstone, the naturalist who happened to own the Hackness estate, took on a competent manager who conveniently doubled up as one of Britain's top geologists. As land steward on the estate, William Smith spent 'six of the calmest and happiest years of his declining life'. The Geological Society got rid of its old president and not only invited Smith to join, it presented him with its first ever Wollaston Medal. (His nephew John Phillips was awarded the same medal 14 years later.) And in 1832, he was granted a government pension of £100 a year.

An ancestor of my own held down that same job as Land Steward at Hackness. Great-great grandpa must have been many years later than William Smith as Robert Turnbull was only four generations ago and Smith was way back in the early nineteenth century.

The refurbished Rotunda isn't entirely nineteenth-century. Its displays are uncluttered, and surrounded with interpretation captions. Among the squares of cardboard, I found my ancestor named as a 'disciple' of William Smith. G-G Grandpa, who turns out to be Smith's successor in the Hackness job, wrote *Index of British Plants According to the London Catalogue*. This is a list of plants that resembles the telephone book, but without the phone book's human drama. It must be interesting to people who are interested in such stuff, as it went through eight editions.

Robert Turnbull, land steward, seems like William Smith to have clambered up the social scale by way of natural history. He sent his clever son William to Smith's nephew Professor Phillips at Oxford, to see if geology could be his thing: but he actually became a Cambridge mathematician. Like many academic Victorian gentlemen, William (a schools inspector in Sheffield) took up hillwalking.

His son, Prof. HW Turnbull, retired to Grasmere for the fellwalking and the climbs. HWT's son was christened Derwent, a name which 88 years later he still dislikes, not after the lake of Borrowdale but the river near Scarborough. Derwent took his own boy up Middlefell Buttress, Great Langdale, in 1961. And that boy started looking at the rocks underneath his fingers, wondering why they were so different from each other and where they all came from.

geological temple, circular in section, with rocks and fossils displayed according to the rational scheme of Mr Smith. The shelves were to be colour coded, with the oldest strata on the lowest levels and the chalk against the ceiling, just below the coastal diorama painted by young John Phillips. The shelves were even to be sloped in the same way as the strata, though that particular idea was (if you'll excuse the pun) shelved.

A local naturalist, Sir John Johnstone, happened to own nearby Hackness Estate and its quarry, and it was he who donated the fine-grained Jurassic sandstone. In 1829 it was finished. Afterwards the Philosophers sat down to devour a reproduction of their new museum made in cake.

Scarborough Rotunda, the world's first geological museum, designed by William Smith

Stones with spots

There are a dozen ways for rocks to be spotty. This is fewer than the sorts of British butterfly (59) so it ought to be easy. But it isn't. One could compile an entire handbook of spotty rocks: concretions of chert, of ironstone, of flint; the various textures of conglomerate; the abundance of odd crystals.

Here is a rundown of rocks with spots in. This speckled pebble may be not sedimentary gravel at all, but little round fossils. Or are the little round fossils perhaps feldspar crystals? Is this chert a xenolith?

If you've worked out whether the rock is sedimentary, volcanic or metamorphic, that helps with identifying the spots. And vice versa. Smaller speckles are best looked at through a lens.

Sedimentary rocks with spots

See also gravel in sandstone (page 37); wormholes (bioturbation) (page 37); colonial coral (page 42); chalk breccia (page 98); flint (concretion) (page 104); calcite cemented beach breccia (page 108); Devil's Toenails fossils resembling concretions (page 139); ironstone concretions (page 148); oolitic ironstone (page 149); ironstone concretions in Old Red Sandstone (page 168); coral (page 178); crinoid limestone (page 179); worm burrows, Marloes Bay (page 211); Silurian fossil fragments (page 211).

ABOVE AND FAR RIGHT
Sandstone conglomerate.
The one in close-up contains angular fragments and also rounded, water-borne ones.

RIGHT Colonial coral

BELOW Ironstone nodules, unusually glossy and angular: longer one 10 cm

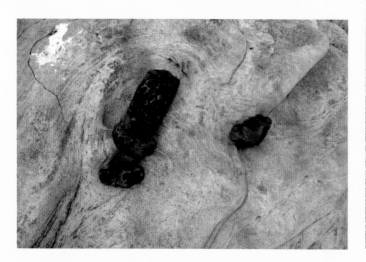

Igneous rocks with spots

See also; crystalline syenite and granite (page 27); xenolith, Ayrshire coast (page 27); Giant's Eye (page 53); vent agglomerate (page 61); crystal gabbro (page 71); pumice bombs (page 77); crystal granite and gabbro (page 79); feldspar crystals in rhyolite (page 79); basalt tuff, and amygdales (page 69); crystalline Larvikite (page 154); Precambrian tuffs (page 216).

ABOVE On the left, basalt block tuff: pumice fragments bound together with volcanic ash. On the right, gas-bubble holes in the lava have refilled, soon after the rock was laid down, with white calcite mineral. The white calcite spots are 'amygdales'.

TOP RIGHT Granite lump, about 40 cm – you can just see the black speckles. The large spots are fragments of the magma chamber wall, incorporated into the granite magma and partly melted: 'xenoliths'.

RIGHT A mass of crystals, so it's igneous and from deep underground. Pale colour, so of the granite family, but without granite's black specks. It's an obscure rock called trondhjemite, at Girvan, Ayrshire.

FAR RIGHT Rhyolite lava, but it's been hanging around in the cooling magma chamber long enough for feldspar crystals to form. A fine-grained rock with individual crystals swimming in it is called porphyry.

Metamorphic rocks with spots

See also magnetite crystal (page 88); mica schist (page 90); fault breccia (page 125); strange textures of serpentine (Chapter 14); Man of War gneiss (page 223).

RIGHT Crystals can also form under the heat and pressure of metamorphism. Garnet schist from the Scottish Highlands.

FAR RIGHT Breccia is conglomerate, but with angular fragments rather than rounded ones. However, breccia can also be formed in a quite different, and non-sedimentary, way by shattering along faultlines.

10. RED SANDSTONE SANDWICH

Old Red Sandstone at Cove Harbour, Berwickshire. The ORS isn't that old: a few miles south, these rocks form the younger, upper layer of James Hutton's 'Abyss of Time' unconformity described in the Introduction.

10. RED SANDSTONE SANDWICH

The Walrus and the Carpenter
Were walking close at hand;
They wept like anything to see
Such quantities of sand:
"If this were only cleared away,"
They said, "it would be grand!"

Lewis Carroll, *Through the Looking-Glass* (1872)

It's traditional to compare rock strata with various forms of cakes and pastry. The three rock systems called Permian, Carboniferous and Devonian form a hamburger, or better still a sandwich – but a sandwich made with real sandstone.

The Permian, all over the UK, Europe and quite a bit of elsewhere, is the New Red Sandstone. The climate was hot, and the scenery was sand dunes and stony desert, not unlike the Sahara. The Sahara is the way it is because of being in the middle of Africa and a long way from any ocean. But its position on the globe, just north of the equator, is also

important. The pattern of winds carries moist air away from it, southwards towards the equator and northwards towards the Mediterranean. So it's strange to think that long ago, global climate was arranged so differently, with hot dry desert up here in what's now the UK.

Strange, but also untrue. As geological eras go by, global winds and climate bands stay almost the same. What changes is our continent, and its place in the world. Three hundred million years ago, in the Permian period, the world was all one big place named Pangaea. Because it was a single huge continent, almost everywhere tended to be a long way from the sea. The bit of Pangaea which would be the UK was Saharan because it was a long way from the sea, but also because it was at the same distance north of the equator as the Sahara is today.

The south continent Gondwana (including South America and Africa) and the north continent Laurasia (including North America, Eurasia, and us) had just joined together to form that supercontinent of Pangaea. So a Himalaya-sized mountain range ran along the crunch zone to the south of us. Like all mountain ranges (when

LEFT Red Sandstone quarry at Marchon, north of St Bees Head, Cumbria. Below, red sand has been laid down in water: these are Red Sandstone shales. Above, desert sand dunes.

RIGHT Red Sandstone at Heads of Ayr. But which is the Old Red and which is the New?

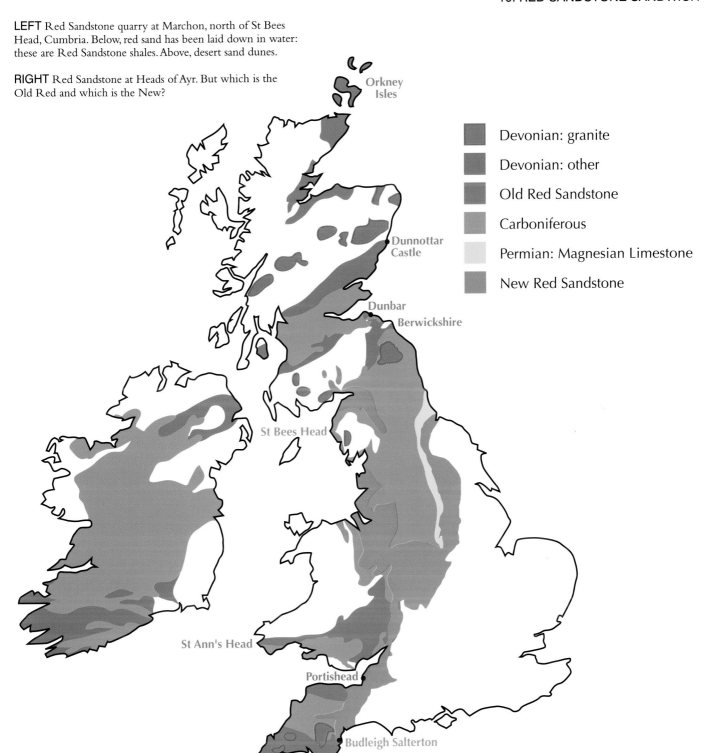

Devonian: granite

Devonian: other

Old Red Sandstone

Carboniferous

Permian: Magnesian Limestone

New Red Sandstone

considered geologically) the Variscans were falling to bits. Storms sent instant rivers roaring down rock wadis and out across the gravel plains of St Bees (Cumbria) and Budleigh Salterton (Devon). In the hot desert winds, the sand kept its bright iron-rust colour, and blew about in big migrating dunes. Red sludgy sediments lay across the floors of flood-water lakes, which then evaporated to pale grey salt pans, as they do in Death Valley, California, today.

Elsewhere in the world, the Permian is divided from the following Triassic by the most devastating extinction of plants and animals in the world's history. It has been called the 'Great Dying'. It may have been a methane hydrate event, a catastrophic release of a chemical normally lurking in ocean sediments. Other explanations? Suggestions include an aster-oid like the one that finished off the dinosaurs 186 million years later – but landing in the ocean, where its crater has

subducted and disappeared; a total stagnation of the world ocean, with all currents stopping and sea life suffocating to death; or a plume of magma from the earth's Mantle, covering much of Siberia in basalt floods, releasing sulphur gases and carbon dioxide. One not usually considered is the possibility that some agile saurian might have evolved a runaway brain, covered the world in a millimetre stratum of compressed lizardly artefacts that we haven't found yet, and wrecked the planet, fortunately extinguishing itself along with everything else.

During this disaster, global temperatures rose by 6°C. The *Glossopteris* flora of Gondwana was destroyed, and the Australian coalfields top out abruptly at 251 Ma. In the ocean, where it's easiest to count fossils, about 98 per cent of species were lost, including the two main sorts of Carboniferous coral. After 300 million years of crawling about, it was curtains for the trilobites.

But here in Britain, the world drifted on unaware. Hot, dry, red Permian desert was followed by dry, red Triassic desert that was for a while (supposing someone was measuring) just 6° hotter still. The two periods are even lumped together as the Permo-Triassic. Or identified in the old fashioned style as the 'New Red Sandstone'.

On a June evening after a day on the Lakeland fells with a haze lying over the sea, it wasn't too hard to think 'hot desert'. The UK's New Red Sandstone seaside is at the western end of Devon's Jurassic Coast. It's also at St Bees, on the sea coast of Cumbria. Here it marks the start of Wainwright's Coast-to-Coast walk, which will end, 14 days later and 100 million years further up the geological table, at Robin Hood's Bay in Yorkshire. The red rock is blotched with off-white. This is nothing to do with desert salt pans, and everything to do with the several thousand guillemots nesting on the ledges. The red rock does have narrow grey-

green stripes in it, which are down to the desert's evaporating salt lakes. Gypsum, alabaster and anhydrite were once quarried out of the grey-green stripes.

Sea-pinks clump under the red-brown rock in a violent colour scheme which no gardener but only great Nature itself would be brave enough for. Yellow eggs-and-bacon (or bird's-foot trefoil) complete the fauvist colour scheme.

The beach shingle at Fleswick Bay contains semi-precious carnelians. These are nothing to do with the New Red Sandstone overhead, but have been brought south by the North Sea Glacier. Apart from gem-collectors, the bay appeals to lovers of the wild who don't mind a bit of a hike.

'Why are you taking all those pictures?' the four-year-old beach appreciator wanted to know. I said I found the rock formations rather beautiful.

'That's just silly,' she told me. Any photos being taken here should be of her. Her parent offered a barbecued chicken leg, complemented with cannabis. I enjoyed the chicken, and the driftwood smoke drifting under the cliff. But with a three-mile walk out, and needing to be quick because of nightfall, I turned down the drugs.

I still think I was right about the rocks being beautiful. The red sand is bound together with iron oxide, otherwise known as rust. Calcite (as in what we humans call cement

168

ABOVE Saltom Colliery, the first one ever to mine coal from beneath the sea. Started in 1729, it eventually reached out for 2 km beyond the shoreline. Horsetail (left, foreground) is a plant almost unchanged, apart from shrinking down from being 10 m tall, since it helped create coal in the Carboniferous.

FOLLOWING PAGES New Red Sandstone at Fleswick Bay, St Bees Head

mortar) and silica are both better cement. Rust, whatever motor mechanics may think about it, is nature's weakest rock glue. Wind and rain wash away the corners of the sandstone, leaving shapes like Henry Moore sculptures but in a jollier colour. Erosion picks out the cross-bedding: the diagonal patterns made by sandbanks underwater, or the advancing edges of a dune.

Even human sculptors find this sandstone fun. It splits along bedding planes, which is convenient; but the beds themselves carve conveniently in any direction, allowing all the flamboyance of the abbey doorway back at St Bees.

Northwards, at and around Whitehaven, are older rocks of the Carboniferous period. The Triassic is New Red Sandstone, and the Carboniferous is coal; except that the Carboniferous isn't so convenient, being also the Mountain Limestone and the Millstone Grit, and various yellow sorts of sandstone.

Some of the sandstone layers in the Carboniferous are not unlike the red desert sand of the New Red Sandstone. But there's an important difference. The NRS is lifeless – unless you're lucky and come across a dinosaur footprint. (The dinosaur cleared off quickly; however big and slavering one's mighty jaws, there just isn't anything out in the red desert to eat.) The Carboniferous rocks, by contrast, feel lived in. Where they're sandy, the sandstone is interrupted by sludgy layers of organic matter. Here and there comes a stripe of black coal; or a red crumble of what was formerly fertile soil. In between are layers of limestone, sometimes coral-white, sometimes lumpy with seashells, sometimes brown with mud.

So as you head north towards Whitehaven, the proud red sandstones end. The ground steps down, and cliffs are lower, and loose. You're warned not to walk underneath them, or

too close to the top. And in case 'feeling lived in' is too subtle a sign, the disused collieries – one from 1729 at sea level, another preserved as a museum behind the clifftops – are a pretty good clue too.

The start of the chapter suggested a sandwich. A sandwich (unless you're Swedish) is two layers of plain bread, lying above and below some tasty organic matter in the form of fish paste, ham, lettuce or cheese. Or, where the sandwich is this sandstone one, the compressed salad is coal, and fossil seafood. The underneath layer of the sandwich is a second slab of sandstone. It's red, and it derives from a different, earlier, desert. It's known with pleasing straightforwardness as the Old Red Sandstone.

Cut two bread slices to make a sandwich: do you turn the upper one over before buttering it, so as to make the two sides of the sandwich match up? Even if you don't mind a ragged sandwich, the New Red and the Old Red Sandstones fit like symmetrical slices off a single loaf.

During the New Red era (the Permian and Triassic) our part of Europe was drifting northwards through the northern desert zone, just north of the Equator, now occupied by the Sahara. During the Old Red era (the Devonian) we were drifting through the corresponding zone in the south, now occupied by Namibia and the Atacama Desert of South America.

During the Permo-Triassic, a big mountain range to the south was crumbling into sand and gravel washed out down desert wadis. During the Devonian, the mountain range was to the north, the consequence of Scotland (plus part of North America) having just crashed into England (and the rest of Europe). Flowing northwards into the Triassic red desert or southwards into the Devonian one, the desert wadis, the sand and gravel, were the same.

In between the Old Red and the New Red, we passed northwards across the Equator. Today that's rain forest and mangroves. The corresponding equatorial hot swamp of the Carboniferous made the juicy salad filling of the sandstone sandwich.

If you're walking along the tops of low, lumpy-sided, bright red seacliffs, it can be hard to decide whether this particular Red Sandstone is the New stuff or the Old. You could search for dinosaur footprints. Find some, and you're in the Triassic era of the New Red Sandstone. Fail to find any, and you're on the Old Red, or the earlier, Permian part of the New Red. Or you could simply have not been looking for long enough.

There's a simpler way, which is to study the small-scale geological map. The New Red top slice and the Old Red bottom slice are well separated; the whole of the Carboniferous sandwich-filling has to come between. The best sandwiches are cut right across to show the delicious fillings. South Wales, Northumberland and Fife put the best-laid sandwich bar to shame. Their tempting range of limestones, sandstones, sea life and plants is laid out in the following chapter.

Portishead: Where old meets new in Red Sandstone

The New Red Sandstone and the Old are separated by 150 million years and the Carboniferous Period. But at one point on the UK shoreline the two manage to meet. That point is Portishead. The distant rocks, beneath Portishead's outdoor swimming pool, show fragments of crinoid (sea lilies) and are Carboniferous. In the foreground, that Carboniferous has been squeezed out. Old Red Sandstone lies below a coarse lumpy conglomerate that's the bottom of the NRS.

ABOVE LEFT The slanting angle of the lower ORS rocks indicates the unconformity: the whole of the Carboniferous Period is missing, and during that time the lower rocks have been tilted to the angle we see now, and eroded flat across the top.

LEFT Stream-type crossbedding in the Old Red Sandstone

BELOW The NRS conglomerate is the first, fiercest flash flood coming down out of the newly formed mountains. It contains lumps of broken-off ORS.

RIGHT The NRS conglomerate trickles down into the eroded gaps in the underlying ORS.

11. COAL COASTS

Three Cliffs Bay is officially Britain's most beautiful beach. The headlands either side are Mountain Limestone. On the right, the hill Cefn Bryn and the back of the bay, both heather covered, are Old Red Sandstone. In the bay is a circular pattern of refracting waves arriving over the sandbar. The intersecting pattern from the left is mystifying until you work out there's a sea arch through the headland.

11. COAL COASTS

'Crab: A small red fish which walks backwards.' Before publishing its dictionary, the Académie Française showed bits of it to mollusc man Georges Cuvier. Cuvier wrote back: 'Your definition, gentlemen, would be perfect, only for three exceptions. The crab is not a fish, it is not red and it does not walk backwards.'

The Old Red Sandstone isn't like Cuvier's crab. In many – even most – of the places where you see it, it is old; it is red; and it is sandstone. The Cretaceous includes the Greensand and the Gault Clay. But it is also, spectacularly, the chalk.

Carboniferous means coal. And the Carboniferous period does contain coal. But mostly it's mudstone and shale. It's the sandy sort of limestone, and the limey sort of sandstone. It also has various volcanoes.

There's one thing that links all the Carboniferous rocks, whether in south Wales, on the bleak coast of Northumberland, or underneath the red-roofed fishing villages of Fife. We felt it at Whitehaven in the previous chapter: but it is no more than a feeling. The Carboniferous sandstones and limestones feel lived-in.

You can tell a house that's been lying empty. Footfalls echoes across the wooden floor. The plasterwork is clean and smooth. There aren't any shades around the light bulbs, and it smells of paint. A place that's been used and occupied, though, has a stair carpet worn down in the middle, scuff marks around the door handles, a scrap of macaroni in the corner of a shelf. The paintwork may be fresh, but there's a layer of dirt underneath it.

This boulder has black bits of twig in it; this rock stratum is all scrambled by something that crawled about in it. Look at this gritstone through a lens, and its speckles are broken shells. The sandstone cliffs, no matter how the sea scrubs the surface, still have streaks of organic muck running through them. These stones, as I said, feel lived-in.

Feelings can let you down, especially in North Devon, where the Carboniferous, like one of those girls you knew at college, is convincing you that it's a place about 100 million years older. Or in Northumberland, it's like those chaps you also knew at college, pretending to be about two thirds of its actual age. You can tell Northumberland's not actually Jurassic, as not one of the tourist leaflets mentions a dinosaur. (The monster frogs of the Carboniferous don't quite cut it like a Brontosaurus.)

A feeling is not a fact. Find out you're on Carboniferous from the geological map – find out by identifying the rocks beneath as Old, rather than New, Red Sandstone. Then slide about over its brownish and greyish and occasionally reddish boulders. Yes, in some indefinable way, it feels lived in.

And the indefinable becomes definite when you spot that the lump on the rock is actually coral. The blue mussel shell is lying beside one the same shape as itself, but a fossil. The beach stone you're holding has a little white disc, a stalk section from a sea lily. (That's unless the beach you're

LEFT Carboniferous coastline: coal in sandstone east of St Monans, Fife

MIDDLE Carboniferous coastline: river-delta sandstone. The Fell Sandstone of Northumberland, not the 'Millstone Grit' delta but its neighbour to the north

RIGHT Carboniferous coastline: Mountain Limestone, as Humphrey Head looks down on the mud of Morecambe Bay

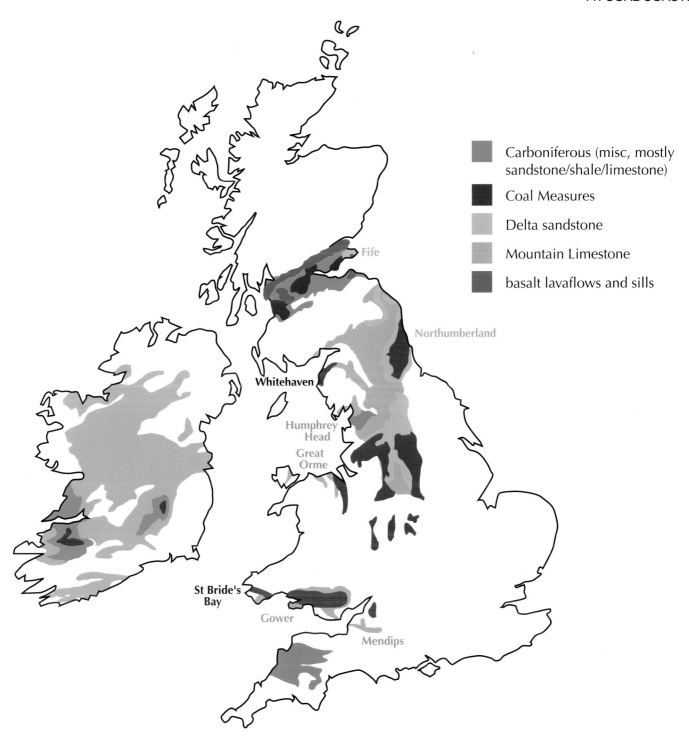

Carboniferous (misc, mostly sandstone/shale/limestone)

Coal Measures

Delta sandstone

Mountain Limestone

basalt lavaflows and sills

standing on is Lindisfarne, where the white discs are known as 'St Cuthbert's Beads', from the saint's rosary that dropped off its string 350 million years ago.) The multiplicity of life, the crinoids and the corals, the shellfish and the worms, all happily ecologising – and a half mile along the shore-line, here's a completely new multiplicity from 10 million years later on. Just suppose Humanity had lived back in the Carboniferous. What a thin slice we would now represent, spread over our mere five million years. You'd walk right past us on your way to buy a Welsh potato pie.

Using the old system, layers of rock in the Carboniferous are named after rocks. On top are the Coal Measures, below

them the Millstone Grit, and at the bottom the wonderful white stuff called the Mountain Limestone. (Using the modern system, the layers are named after places in Belgium.)

The Coal Measures were described in the Red Sandstone chapter. They are sea-coast sediments; but 'Carbonivia', at least in the coal age, was not a fun place for your seaside holiday. Blue seas under a tropical sun give a limestone sea bottom. But great river deltas swished mud and silt over that limestone, and as the delta rubble built up from the shoreline it added a layer of sand and pebbles. As the gravel bed built up to the sea surface, swampy forests grew, with tree ferns and stringy horsetail.

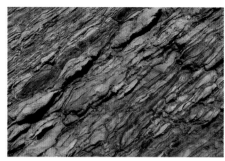

LEFT Two small coal measures, about 10 cm thick: Crail, Fife coast

MIDDLE The typical Carboniferous rock is 'all of the above'. An ebullient mix of sandstone and mudstone: this one also has a 5 cm band of ironstone.

RIGHT Impression of tree, with thin coating of coal: Crail, Fife

BELOW Skilful geologists can determine a fossil's exact species, and date its rock within a few million years. The less sophisticated sometimes get to pin down a distinctive fossil group or family into a specific geological period. Thus the Devil's Toenails indicate the Lias (Chapter 8); and this rugose coral is a marker for the Carboniferous.

BOTTOM A tabulate coral, the simpler of the two sorts that flourished in the Carboniferous. All the tabulate and rugose corals died out in the great end-Permian extinction. Modern corals are of a third kind called 'Scleractinian'.

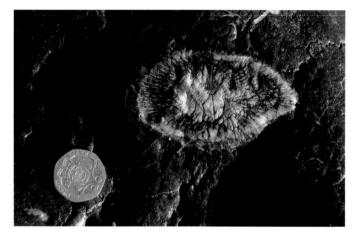

Giant amphibians, frogs the size of crocodiles, splashed in the ooze between the treetrunks. Dragonflies the size of seagulls hovered in the shafts of sunlight. No birds sang, as birds had yet to occur – that is precisely why the dragonfly was able to adopt the bird size and lifestyle. A millipede 1.2 metres long tiptoed across a sandbank on the shoreline of what would one day be Fife; an hour later it tiptoed back the other way.

Then the sea level rose, or else the sea-bottom sank. The river re-routed itself and formed a new delta. And today we see the sequence of limestone, mudstone, siltstone and coal, repeated again and again.

The coal coasts of 300 million years ago were hot, swampy and smelly. And the coal coasts of now are not the loveliest in the land. The northwest coast of Cumbria is crumbly like cheap brickwork. The coal coast of Northumberland can be described as dingy. St Bride's Bay is high, and yellow stained with black, and collapsing.

The Fife coast too is backed by greyblack coal measures. So how come it's so sweet? It could be Fife's Carboniferous sandstone. Its rich ochre tones ripen to warm orange, as Carboniferous bacteria got happy in a mud layer and accumulated atoms of iron. The orange stone rises in trim, weatherproof cottages; and the same iron colour, just this side of stridency, is on the pantiles overhead.

Seaside Fife is the curve of stone harbour walls, mimicking the curved strata along the shoreline. It's the bright paint chosen by the villagers of Pittenweem and St Monans over hundreds of years. It's the occasional dolerite intrusion giving a upward jerk to the shoreline. It's the economics of Fife being not far from Edinburgh and Scotland's first capital at St Andrews, so the well-off Fifers were willing to pay for their fish, with the fish in turn paying for the handsome stonework of the harbour.

When they asked a seventeeth-century Scotsman if he'd been abroad, he said he knew a man who once went to Crail. Doesn't this story just say Crail was an important place to go to? Envious of Crail's twelfth-century sandstone kirk, Satan flung a stone at it from the Isle of May. An angel deflected his aim, and the missile landed at the entrance gate, 30 yards short. There it lies today, a lump of black volcanic rock. And as the Isle of May is detectable from 10 miles away as dolerite sill, we know the story must be true.

Crail's yellow rocks are river-delta sediments. They're wrinkly where the water squeezed out, and diagonally striped with cross-bedding. As in Yorkshire's Jurassic, the river that

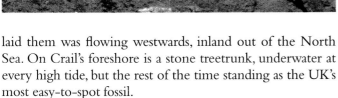

laid them was flowing westwards, inland out of the North Sea. On Crail's foreshore is a stone treetrunk, underwater at every high tide, but the rest of the time standing as the UK's most easy-to-spot fossil.

Below the coal, the Millstone Grit is another sandstone, brownish, and cross-bedded in the style of another river delta. That river was big: its delta stretches from North Yorkshire right down along the Pennines to the Peak District. But it didn't reach what today is England's coast. So the 'Millstone Grit' age rocks around the shoreline aren't actually made of Millstone Grit – and perhaps the noncommittal modern name of 'Namurian' is preferable.

LEFT Brachiopod, about 10 cm wide

RIGHT Crinoids (sea lilies) are fairly common in the Carboniferous, less so in other periods such as the Jurassic.

BELOW Track of 500 of the feet of giant millipede, 1.2 m long – the prints of the other set of feet are off the edge of the slab. Crail, about 50 m west of the fossil treestump.

Seen on the shoreline, these 'Millstone Grit age' (or Namurian) rocks do include river-delta sands. But also, they offer more of the muddy limestone, limey siltstone, and black swampy bits; the Coal Measures but without the coal.

But below the Millstone Grit is a rock that I refuse to call the 'Dinantian'. It's Mountain Limestone by (old-style) name and Mountain Limestone by nature. White or pale grey, it's made up of coral reef, and of shallow-water limestone from the blue seas between the reefs. It stretches across the Yorkshire Dales, forming the strange scenery of limestone pavement, and even stranger underground scenery for the Yorkshire cavers.

At Humphrey Head, on the southern shore of Cumbria, that limestone pavement runs down into Morecambe Bay. Oak trees sprout like house-leeks from the off-white crag, and above them is half a cavern, broken open by the sea. Weston-super-Mare has its pier with funfair and slot machines. But if you stroll southwards along the long beach, there's Mountain Limestone where the Mendips meet the sea, with the final foreshore rock composed of crinoid bits.

LEFT Crail harbour, and its fossil tree

ABOVE Isle of May

BELOW Lang Shank, east of St Monans, Fife. Dolerite is an intrusive volcanic rock that has nothing at all to do with dolomite, even though that too is found at St Monans. A few miles offshore from St Monans, the Isle of May shows the columnar jointing and flat top of a dolerite sill.

The Mountain Limestone juts out from the holiday coast of north Wales as a great offshore lump north of Conwy Castle. Pen-y-Gogarth is in English the Great Orme, except that Orme is not English but Norse and means a 'worm' or sea-serpent. Conwy has tamed the snake, with a winding road, a nine-hole golf course and a cable car that lifts you to the car park that constitutes its summit. There is also a hundred-year-old tramcar ride.

In Norse legend, the great Midgard Serpent winds right around the world. So it's no surprise to find this limestone sea-snake again at the other end of Wales.

Gower

> Time held me green and dying
> Though I sang in my chains like the sea.
>
> *Fern Hill*, Dylan Thomas

Worm's Head is the extreme tip of Gower. At least, it's the tip of Gower for five hours around low tide. The rest of the time it's an island. It's a tilted, cave-pierced, wonderful lump of Mountain Limestone, reached by half a mile of scrambling over rocks and puddles.

Gower is Officially Beautiful. It's been beautiful ever since 1956, when it was the UK's first ever designated Area of Outstanding Natural Beauty. It's been praised by Dylan Thomas, who grew up on its heaths, fell in love on its beaches, and wrote *Under Milk Wood*. In a 2006 BBC poll, Three Cliffs Bay was voted the UK's best beach. OK, so most of those voting had been to busy Three Cliffs Bay but not to Sandwood Bay up next to Cape Wrath (see page 18). Three Cliffs could be the top beach spot for those of us who like our beauty draped with beach towels and people in bikinis.

If Three Cliffs is the tops, there are three reasons for that, and all three reasons are to do with the rocks.

First and fundamentally, it's the Mountain Limestone. Gower is where the Mountain Limestone makes seaside scenery not just over one headland but a whole coast. Limestone's golden colour is a good start. Limestone is hard; it makes strong shapes against the sky. Limestone dissolves in seawater to form striking holes and rounded edges, sea arches and blowholes.

Limestone dissolves in rainwater, and limestone soils are thin. But limestone is rich in calcium and other minerals, so the thin soils are fertile. No heavy mud clogs the clifftops, but bright green rabbit-cropped grass, with wild flowers.

Mountain Limestone's home is in the mountains of Yorkshire and the Peak District. There it lies flat, as low crags called scars and the occasional limestone pavement. But Gower is further south, and lies in the crumple zone of the Variscan mountain building. Gower cliffs are slanted and folded, jagged and broken about; and that's the second geological reason for Gower.

The third rock reason is negative. North of Gower are the Coal Measures. To get to Gower, you travel the coal country, and enter through the coal port of Swansea. A place could be a lot less attractive than Gower and still look great after Port Talbot.

The limestone in the Yorkshire Dales is easy to slip over on. But it's in drizzly Yorkshire that it grows the coating of algae. Limestone doesn't have to be like that. The limestone mountains of Europe are rough enough to leave your fingers sore, and grip the bootsole like Velcro. Salty, moisture-stealing sea winds are as unfriendly to algae as the snows and scorching sun of northern Spain. Scrambling the shoreline of Tears Point, on glaring rough limestone under a hot afternoon sun, I suddenly felt myself where I had been two summers before, on bouldery sea-slopes of Mallorca.

Back in the Carboniferous, the Mediterranean was an ocean called Tethys, dividing the world in two. Thick layers of limestone formed along its warm northern edges. The inexorable arrival of Africa is crushing and folding that limestone, and raising it into the air to form our favourite holiday islands.

Tough, rough, white limestone, raised and folded by a continental collision? If we ignore the temperature and the colour of the sea and the sky, Gower is the UK's Greek island.

The Carboniferous: not quite so unique as it ought to be

The Carboniferous is when the UK was passing through the rich green equator, in between the two times of red deserts. The Carboniferous has tropical reefs formed from corals,

and coastal swamps compressed into coal. It has huge beds of limestone, and distinctive alternations of limestone and sandstone and shale. This makes the Carboniferous unique.

Or at least, it ought to be. Except that the Jurassic period, 150 million years later, also has corals and crinoids. The Jurassic period has jet, and jet is a form of coal. And the Jurassic period has even more distinctive stripy-layered rocks of limestone and shale.

In the Lias rocks of the Jurassic (Chapter 7) we saw clear, limestone seas alternating with muddy ones and river deltas. In the Carboniferous the same stripy layers happen again, and are called 'cyclothems'. As in the Jurassic, there can be up to 100 of them, one on top of the next. The top layer of each cyclothem is where the coal is. This rhythmic tick-tocking of the seabed requires the same sort of explanation as it did in Chapter 8's Lias layers.

The cyclothem layers, just like the Lias ones, require the local sea level to rise for each muddy layer, every hundred thousand years or so, and then sink back down again. But the layers will soon fill up the sea. So there needs to be also a slower and more gradual rise that allows each new peak

ABOVE Gower, looking back across the tidal causeway from Worm's Head. On the left, a great arch in the rocks (anticline) brings up the older and lower Old Red Sandstone, in the heathery Cefn Bryn. The softer ORS has allowed the sea to carve the great Rhossili Bay, whose further edge, three miles off the left edge of the picture, is another limestone headland, Burry Holms.

RIGHT Gower's different ways to make bays.
At Rhossili Bay, an upward arch, or anticline, in the tough limestone lets the sea get through into the Old Red Sandstone underneath. (See also Robin Hood's Bay, Chapter 9.)
At Oxwich Bay, a downward sink, or syncline, lets the sea get over the limestone into the softer Coal Measures above.
At Three Cliffs Bay, the limestone tilts steeply upwards towards the land. The sea has broken a gap, to form a wide, round bay in the Old Red Sandstone behind. (See also Mupe Bay and Lulworth Cove, Chapter 8.)

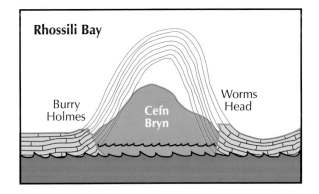

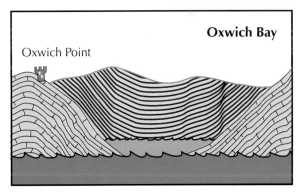

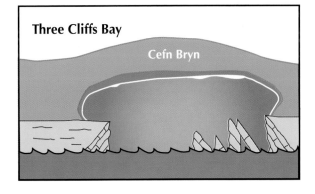

Mountain Limestone at Tears Point, Gower

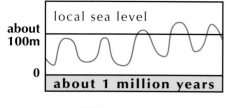

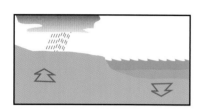

in the local sea level to be a few metres higher up the beach than the one 100,000 years ago (see above).

The up-and-down bounce is explained by an ice age and the 100,000 year Milancovitch climate cycle described in Chapter 8. When the world gets colder, water out of the ocean falls as Arctic snow and builds up into icecaps. In the Jurassic, that ice-age suggestion was a tentative one. But we know that there was, indeed, an ice age in the Carboniferous. Chapter 5 described how the great Gondwana continent lay over the South Pole, and the scratch marks of its moving ice are there to be seen in the Carboniferous rocks of Africa, South America and Australia.

The underlying gradual rise could be interpreted as a general rise in the level of the world ocean – except that, during the later Carboniferous, the world ocean was gradually falling. Instead, then, it has to be a fall in the local sea bed. As erosion wears down mountains and deposits them on the sea bed, the sea bed gets heavier and the land gets lighter. The continent is floating on the mantle below, so as the sea bed gets heavier, it sinks. As the land gets lighter, it rises. And as it rises, it exposes itself to erosion, so that more rain and rivers can carry it down into the sea.

The Carboniferous and Jurassic periods ought to be quite different. They are 150 million years apart; and given the UK's northwards drift, that also makes them 1,000 miles apart, in two quite different climate zones. There are two main reasons why they are actually quite alike. The first is sea levels. Both were times when the world ocean was rather high. Much of Europe and other continents wasn't dry land at all, but shallow sea with rivers running into it. Hence delta sandstone, limestone, shale, and seashells aplenty.

The other reason is a quirk of world climate. The coal times were hot here in the UK because we were on the equator. But the Jurassic was a time of warming world-wide, with overall temperatures up by about 7°C. The Jurassic UK was drifting north through what we'd now call the Mediterranean

ABOVE Limestone and shale layers at Craster, Northumberland. Behind, the Great Whin Sill marches out to sea. The upright columnar jointing of the dolerite is at odds with the level bedding of the sedimentary rocks.

RIGHT Limestone sea stack: but the top half is 150 million years younger than the bottom. West of Dunraven Bay, Glamorgan.

regions, but its climate was hotter than that. Thus the Jurassic, like the Carboniferous, has rocks made of coral.

In Glamorgan, the New Red Sandstone is mostly missing, and the Carboniferous and the Jurassic lie up against each other. The Gower, in West Glamorgan, is Mountain Limestone: Traeth Mawr, in East Glamorgan, has limestone and shale of the Jurassic Lias. And between the two, at Ogmore-by-Sea, the two meet on one beach. Or rather, on three beaches … the present day beach of Ogmore is Beach Three. Above the sand, Jurassic coral grows out of what was then an island edge, and that island edge is Beach Two. The island the Jurassic coral grew on was formed of Carboniferous coral; that coral when still alive was the first and bottom beach, Beach One.

Which is which? The fossils are alike, because of the similar climates, but at the same time pretty different. Both Jurassic and Carboniferous have seashells, and you need a practised eye to tell a brachiopod (Carboniferous) from a clam (Jurassic). But the Carboniferous has lots of corals, including the spectacular rugose corals typical of the Mountain Limestone. 'Rugose' refers to the ridged surface, but these big corals almost are the roses of the rocks. During the Great Dying in the middle of the New Red Sandstone time, that sort of coral was completely killed off, along with most of the other sea creatures. You don't need to distinguish each species of sea-rubbed fossil. That big-thumbed, rugose class of coral says Carboniferous as clearly as a bun penny says Queen Victoria.

The Carboniferous has crinoids, often making entire beds out of the bits – but the Jurassic has a few crinoids too.

After the Great Dying, the two surviving ammonoids flourished and multiplied, becoming the true ammonites and occupying ocean lifestyles so conveniently cleared by the disaster. The Carboniferous has occasional small ammonoids. But where you find a nice big ammonite, you're likely to be in the Jurassic.

Tracking rocks by their fossils is the technically correct way to do it. The rugose corals mean that when it comes to the Carboniferous, and without ten years practice peering at stones through a lens, you actually can. Or else, you could head east for a mile or two, where it becomes altogether obvious.

Below the caves of Southerndown, bedded limestone of Carboniferous age lies under bedded limestone that's Jurassic. The same stuff above and below – but with a crucial clue. The Carboniferous rocks were formed before the great Variscan mountain-building, and the Jurassic ones after it was all over. The Jurassic rocks lie flat and undisturbed, while the Carboniferous ones are bent and tilted. Result: a classic unconformity. Level beds of Jurassic limestone lie across the eroded-off ends of slanted Carboniferous beds below. The angle change is as clear as at Siccar Point – even if it does lack Siccar Point's convenient colour coding.

This uncommon unconformity is only seen at low tide. To reach it from the east, you have to scramble around a rocky corner. Stand and marvel for too many minutes and the tide could cut you off. But the time-gap sea stack is worth getting your feet wet.

When William Smith traced all the different layers of the rocks, the holes he went down were digging for Carboniferous rocks. Later, he was fossil-finding in the cuttings for canals built to carry the dug-up Carboniferous rocks. People in those days, even rich ones, were interested in everything just so long as it was interesting. But also because Smith made suggestions like: don't dig for coal here, you've 5000 ft of chalk to get through on the way down.

Wegener looked back into the Carboniferous. It was then that tropical Europe and ice-age Africa made the idea of continents standing still even more absurd than continents sliding around. And he traced back the supposed sliding-around to one single landmass, Pangaea, that came together in the Carboniferous.

At the end of the Carboniferous, great piles of plant life had sunk into the coal swamps. All of the carbon in that future coal had been extracted from the atmosphere. The result: a reverse greenhouse effect, and a period of global cooling. As we saw in Chapter 5, central Africa at that time lay across the South Pole. There was plenty of land there on which ice could form. Ice reflects sunlight straight back into space, and this only cools the earth further. The result, at the end of the Carboniferous period, was a major ice age. Enough ice piled up on the Antarctic continent to lower the world's oceans by some 50 m.

At the end of the Carboniferous, the atmosphere was lower in carbon dioxide than it had ever been, and the world was cool. It had never been so cool, and so low in CO_2, before: it would not be so cool again for another 300 million years. The other cool, low-CO_2 time is right now.

And right now, the interesting thing happening to the Earth is the way we're digging up the fossilised tree-ferns and horsetails of the Carboniferous, and turning them back into carbon dioxide. It's an undoing, as it were, of the entire Carboniferous period. At the time of writing, the results of this climate experiment are excitingly unpredictable.

12. CONTROVERSIAL DEVON

Tough, grey, twisted rocks of the West Country: Hole Beach at
Trebarwith Strand, north Cornwall

12. CONTROVERSIAL DEVON

It's a mistake to go back. The rocks may remain – 40 years is less than a moment in geological time. But they update everything else and spoil it.

Except when that place is Crackington Haven. Behind north Cornwall's hedges are the same slate-roofed grey farmhouses, the same tiny fields in the vigorous green of England's southwest, where tree heathers and fuchsias dangle over the harsh stone. The lane is as narrow as it ever was. OK, the road sign on the way down warns of a 20 per cent gradient, but I'll always mentally convert that to the old one in five. I mentally convert every such sign and compare it with the one in five down into Crackington, which is my lifetime standard of a steep little lane.

We'd spread our beach blanket at the left side of the bay, and the grey slabs above are where I used to scramble about. But I was wrong when I said the rocks would not have changed. During the missing 40 years they have managed to get even more interesting. Each great grey bed is one underwater avalanche of sand and mud. The ridged bumps, useful footholds to small sandalled feet, are not fossil ripples on the sandy rock bed. They are something called load casts. As it

came to a stop, the sludge avalanche sagged into the mud below to form these lumps and ridges. The upheavals that cracked up Crackington have turned it right over; the surface I once scrambled over is actually the underside of the slab.

As you eat your boiled egg with its seasoning of sand, each wave comes slightly short of the one before and the beach gets bigger. Until you can walk around the corner to find the tall cave, so slippery and smelling of seaweed. The tide goes out some more, and there's a pool six feet above beach level, cupped between two of the tipped-up rock beds. You can take off your sandals and sit in it – if you're small enough – with the seaweed tickling your toes, as you look out over the busy beach.

No wetsuited show-off surfers here stand on the wavefronts. The beach is too narrow and too crowded. Instead, the half-length surfboard where, with a much smaller amount of skill, you ride the wave lying down. They were made of plywood, curved upwards at the front, and we hired them for sixpence. Today's version, in bright plastic, probably doesn't leave those scratch marks across your stomach, so painful when the salt water gets in.

Just one thing is missing: the black, glossy beach oil that used to lurk among the pebbles. Back then, the Merchant Navy dropped lumps of oil as it went about its shipping business; you went to the bucket-and-spade shop and got some solvent to wash it off. The solvent smelt interesting but didn't work very well, as the special pad under the screw-top lid blackened over. So perhaps today's heightened environmental awareness of oil blobs is a bonus.

Once we went when there was a storm. Daddy was in the Navy, and knew useful stuff about storms. Every seventh wave is an extra big one. When there are whitecaps in the open sea, then it's Force Five. I worked around the rocks on the north side to watch the waves come in, and every seventh one came almost onto my toes. And then what must have been the 49th, or even the 343rd, came right over the top of me – I remember clinging onto the rock, and wondering when I'd have to let go and get swept into the sea. But I didn't let go, so I'm still here 40 years later writing these words.

Some boys of today were scrambling on the Crackington slabs, and a family was sitting on the rocks where I wanted to place my own nostalgic bottom. On the cliff path, pink heather and yellow gorse hunched under the wind. Black billygoats lurked in a valley that ended in empty air above the sea. I headed down to the car park to see if the Cornish pasties had been updated. Maybe it was the struggle against the clifftop breezes, but the twenty-first-century, deep-fried version was even better than the original.

Modern cars find it much easier to manage the one in five on the way out.

LEFT Crackington Haven, north Cornwall, and the tough grey sediments known as greywacke

ABOVE Recumbent fold – one lying on its side – at Crackington Haven

BELOW Sea cave, Crackington Haven. Note load casts everywhere, but clearest just to right of the cave.

The further you get from London, the older and more mysterious are the rocks. The Geological Society's headquarters at Burlington House stands on the London Clay, a sediment of Tertiary age, so young it hasn't had time to harden into proper mudstone. Head west or north, and at Salisbury Plain you're on the chalk. Behind the chalk is the long diagonal stripe of the Jurassic, first the Oolite in its various shapes and then the Lias lying underneath. Beyond, and below, is the New Red Sandstone; below the New Red Sandstone lie the Coal Measures, the Millstone Grit, and the Mountain Limestone. And underneath all that, the browner sandstone of the Old Red – with, if you're clever or lucky, some flattened fossil fish.

By 1831, the year the Geological Society gave William Smith his medal, all of this was making pretty good sense. From Lyme Regis to Brighton, from Yorkshire right up to Edinburgh and Fife, its members were tracing the layers, correlating the fossils, interrogating roadmen and navvies, and refusing to be taken in by half-hidden faultlines and unexpected kinks or bends.

In making sense of it, they were helped by the sequence of Universal Formations. These are shown in William Buckland's imaginary section of the strata of England and Wales (and with any luck, the whole world) as understood in 1832. The universal formations worked in three ways. Firstly, the New Red Sandstone of Exmouth, and the New Red Sandstone of St Bees, just looked like the same rock. In the same way, to an experienced bird-watcher, a dunlin in Devon just looks like a Cumbrian dunlin, even before the binoculars focus onto the slight beak curve or the black patch underneath.

Secondly, there's what came to be called 'stratigraphy'. Trace the layers, and the New Red Sandstone will always lie above the Coal Measures, always underneath the Jurassic Lias. And finally, to confirm it all there are the fossils.

Beyond and below the Old Red, though, things got tricky. Cornwall, Wales and Highland Perthshire were more than a day away by stagecoach. Country parsons eager to show off their fossils and lend out a labourer and a pony – these helpful gentlemen were thin on the ground that far out. And emerging from heather moorland or black peaty bog, the rocks themselves were a whole lot less easy to read. The Lias lay in nice layers, but these rocks from deep back in time were wrenched and twisted. Even the basic bedding was often overwritten by a slaty cleavage at a different angle altogether.

William Smith's insight, that a rock layer could be pinned down by the sea shells and ammonites inside it, was the way to make sense of such tricky stuff. Well, these rocks in their utterly jumbled layers, inconveniently distant from London, were almost entirely without fossils.

Unsurprisingly, gentlemanly geologists were happier nearer home. There was plenty of fun to be had piecing together the muds between the Coal Measures, tracking dinosaurs across Dorset, digging shells out of East Anglia. Equally unsurprisingly, other gentlemanly geologists were eager for the challenge. One such was Roderick Impey Murchison.

Roderick Impey was born in 1792, the orphan son of a wealthy landowner in northern Scotland. He served in the Peninsular War, married a wealthy heiress, and pursued the fox until, despite the heiress, he found himself running out

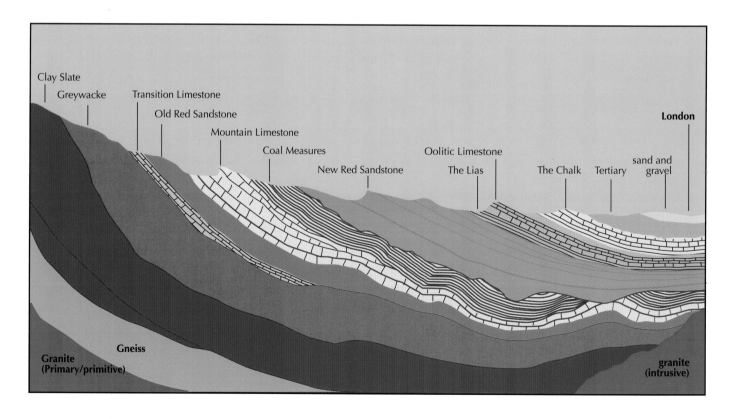

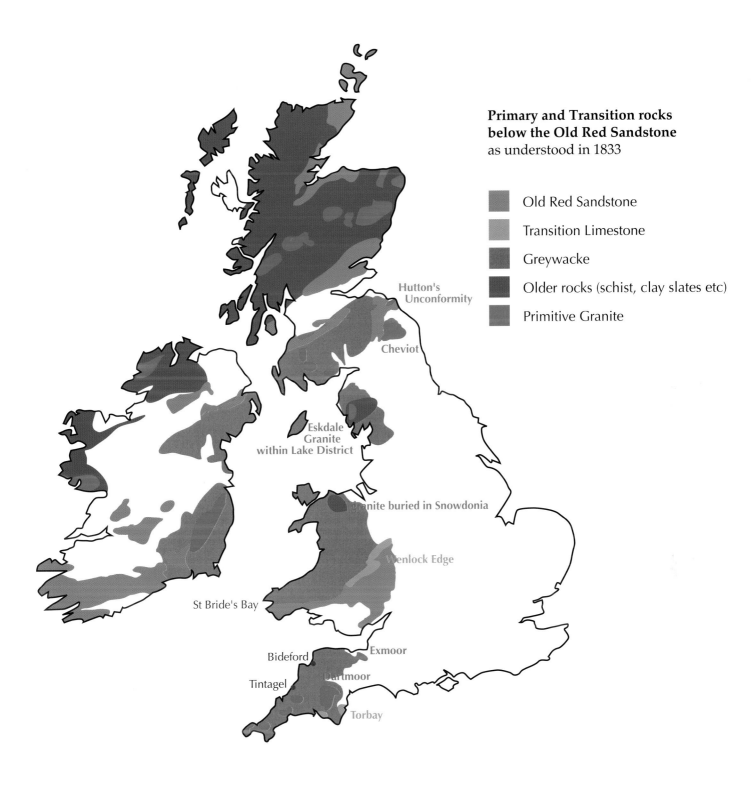

**Primary and Transition rocks
below the Old Red Sandstone**
as understood in 1833

Old Red Sandstone

Transition Limestone

Greywacke

Older rocks (schist, clay slates etc)

Primitive Granite

Hutton's
Unconformity

Cheviot

Eskdale
Granite
within Lake District

Granite buried in Snowdonia

Wenlock Edge

St Bride's Bay

Bideford

Exmoor

Tintagel

Dartmoor

Torbay

LEFT An imaginary section of the strata of England and Wales (and with any luck, the whole world) as understood in 1832. From a diagram by William Buckland, Geology Professor at Oxford. With hindsight, everything in this diagram from the Old Red Sandstone upwards is more or less okay. Everything below is not.

ABOVE The Greywacke and related layers, as understood in 1833. Even then, younger geologists were ceasing to believe in the 'primitive granite', lying under everything else and seen emerging on Dartmoor – they interpreted all granite as being intrusive, squeezing itself up through pre-existing, older rocks. This labelling of the Torbay and Plymouth limestones as Transition would be conjectural.

of cash. Geology might not be so much fun as foxhunting, but it was cheaper. Battle and bloodsports turned out to have been fine training for geological field work.

Among all the bent and complicated rocks, there was one foothold. The ancient, chaotic mountain ranges had their own universal formation, their colour-coded rock type. It was the tough, unlovely sandstone seen at Crackington Haven, and known as Greywacke. Sometimes it made massive beds like reclining elephants. In other places, somewhat thinner grey beds alternated with black shale – when the shale was picked away by the sea, the grey beds stood out in bold stripes. But whether plain or in stripes, it was curved and creased like a tablecloth thrown aside after a party.

The Greywacke lay across Scotland's southern uplands like a great grey doorstep at the entrance to the even older and more intimidating Scottish Highlands. Greywacke surrounded the Lakeland mountains in a rough ring, and rose in a great lump out of central Wales. And a band of Greywacke crossed Devon from the Bristol Channel to the edge of Dartmoor.

The first question: was the Greywacke the next thing down from the Old Red Sandstone? Where the two met, it was mostly at an unconformity. Hutton's Unconformity, explored in this book's Introduction, is precisely an unconformity between the Greywacke (underneath) and the Old Red Sandstone (on top). What the unconformity starkly shows is that between the Greywacke and the ORS, time has passed. Siccar Point, indeed, represents not just time but an 'Abyss of Time'.

If there wasn't anything between the Old Red and the even older Grey, then somewhere there ought to be a smooth, conformable, transition between them. If on the other hand there was something between, then it behoved the geologists to find that something. And on the English–Welsh marches, around Wenlock Edge, Murchison found it. Limestones and shales carried brachiopod shells and coral not seen in any of the existing strata of England. Better still, these 'Transition Limestones' gently gave way to Old Red Sandstones above, and lay gently on grim Greywacke below. Like the family tree of the Welsh Princes, the succession of the strata was unbroken all the way. These 'Transition Limestones' would be the basis of a new geological period, the Silurian, which Murchison would start spreading across Wales, Scotland and the world.

But then, in 1834, Murchison's Greywacke explanations were suddenly contradicted by some 'interesting fossil plants' from Devon.

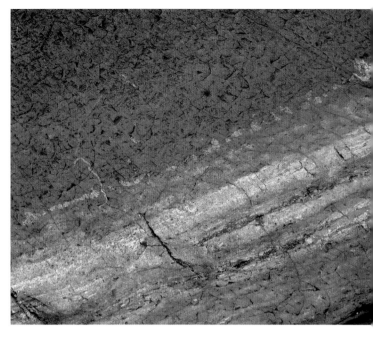

LEFT Greywacke arch, Pwll March north of Newgale on St Bride's Bay, southwest Wales. The rocks are Ordovician.

ABOVE AND BELOW At Marloes Sands, Pembrokeshire, the Old Red Sandstone lies contentedly on top of grey Silurian sandstone.

This, Gentlemen, is my nose.
you, before we saw you, among men without noses, you cannot possibly have a nose.

My dear fellow,' — your account of yourself generally may be very well, but as we have elsewhere

Henry de la Beche's nose

Henry de la Beche was the gentlemanly geologist from Lyme Regis who (as we saw in Chapter 6) was a friend of Mary Anning the fossil-hunter. He should have had the geological name of 'Henry Beach'; the Frenchified surname was an affectation of his grandfather's. Henry was a lively young man, thrown out of the army for insubordination and for being sympathetic to the French Revolution. He had learnt military draftsmanship and mapping, which were useful skills for the future geologist – William Smith himself was a surveyor. But Henry's mischievous pen was also for entertaining Mary Anning with pictures of the plesiosaur; and it would become a weapon against Roderick Impey and the Geological Society.

De la Beche's fossil-hunting was funded by an income of £3,000 a year from his father's sugar plantation in Jamaica. The abolition of slavery in 1833 left him short of funds. He might have to live in Jamaica – and young men from England living in Jamaica didn't usually live for very long. In the nick of time, he picked up a government job making a geological survey of Devon.

For this he was to receive only £300 – equivalent to about £25,000 today. But Devon was straightforward: the place was Greywacke. In the northeast, the Greywacke ran in underneath the New Red Sandstone of the Devonshire farmland. In the south, the Primary Granite, the fundamental bedrock of the whole world, poked up through yellow moorland as the tors of Dartmoor.

An intellectual challenge was supplied by some funny limestone at Berry Head, above Torbay. William Smith's map had made it an outlier of the Mountain Limestone. This would imply that the Old Red Sandstone, which should have lain on the Greywacke, had eroded completely away before the Carboniferous. At which point, Mountain Limestone had been laid directly onto the Greywacke, with the junction being an 'unconformity' because of the missing Old Red. And then the Mountain Limestone had eroded away in its turn, leaving this single lump.

That was one theory of the Torbay limestone. But it would make more sense for it to be part of the Transition Limestone, the new stuff discovered by Murchison lying directly on top of the Greywacke of Wales. The fossils were unforthcoming. Some were of sorts that belonged both to the Mountain and to the Transition Limestone. The others were new and could be assigned to neither.

De la Beche moved north. Around Bideford there were scrappy seams of coal. The chaotic Greywacke rocks that ran from Dartmoor to the Bristol Channel were known, indeed, as the 'Culm Measures'. Culm was a local term for low-grade coal, and 'Culm Measures' was nicely confusable with 'Coal Measures' to anybody who didn't know what was what. But to those who did, a bit of bad coal, way down below the Old Red, wasn't a problem. Not all coal is Carboniferous. There were oddments of coal in the Oolite and Lias (what we'd now call the Jurassic). The jet at Whitby was coal, even if Queen Victoria did treat it as jewellery rather than fuel.

What de la Beche found at Bideford, though, was worse than a bit of coal. He found fossils: horsetails, ferns and tree ferns, all clearly Carboniferous.

This was impossible – and Roderick Impey Murchison had no hesitation in saying so. He professed himself 'astonished that so experienced a geologist should have fallen into so great a mistake—as to fancy that the specimens on the table had anything to do with Transition [i.e. Greywacke] rocks'.

The rocks were on the table, but de la Beche wasn't – he couldn't afford the stagecoach to London. In his absence a magnificent row broke out. Either de la Beche had missed something, or else fossils were not, after all, a guide to rock types. Murchison was unequivocal. The fossils were right, so de la Beche was wrong. The coal and the tree ferns were in a small outlier of the Carboniferous lying *on top* of the Greywacke. De la Beche had blundered. He had failed to spot the unconformity at the base of this coal-bearing Carboniferous oddment, despite having nice clear sea cliffs to spot it in.

De la Beche was upset. Murchison had never even been to Devon. 'As I had toiled day after day, for months in the district, examining every hole and cranny of it, this was a pretty good go of preconceived opinions against facts, which are so plain that the merest infant in geology could make no mistake.' And he illustrated with a cartoon of himself, in a fieldworker's scruffy jacket, confronting the tailcoated theoreticians of the Geological Society.

'This, Gentlemen, is my *Nose*.'

'My dear Fellow! – your account of yourself generally may be very well, but as we have classed you, *before we saw you*, among men *without noses*, you *cannot possibly have a nose*.'

Devon was not the only disaster. At St Bride's Bay in Pembrokeshire, de la Beche had identified Culm Measures like those of Devon, also embedded within the Greywacke, and also with coal-type fossil plants. However, Murchison was able to show that these coal fossils were not within, but on top of, the Greywacke. There was an unconformity: a gap in time. It had been concealed by the general greyness of both Greywacke and Coal, and by the refolding and mangling of the entire area.

LEFT De la Beche's nose, as obvious as the fossil ferns of Bideford. His sketch is on a letter to Adam Sedgwick.

BELOW The rocks of Devon and Cornwall are faulted and folded. Even with clear cliff exposures, it's not straightforward to work out what belongs on top. Greywacke sandstones at Cambeak and Cam Strand, north Cornwall.

Wrong once in St Bride's Bay, de la Beche had undoubtedly made the same mistake again in Devon. In the spring of 1836, Murchison and Adam Sedgwick went down there to sort out the 'little coal hole at Biddeford'. Firstly, De la Beche's strata needed to be redrawn with the problematic Culm Measures on top of everything else. Second, they needed to pin down the unconformity between these Culm, now relabelled as Coal Measures, and the underlying Greywacke. That unconformity would be a big one, representing the missing Mountain Limestone, Old Red Sandstone, and Silurian. It would be shown in sea cliffs on the north coast of Cornwall, to west of the Culm; and then again, at the Culm's eastern edge, on the north coast of Devon. So it would be easy to spot – and it was going to be ever so embarrassing for poor Henry when they did.

The unconformity wasn't obvious at once. They followed the coast from Bideford westwards, and arrived at Tintagel.

TOP This nondescript shoreline at Fremington Quay, near Bideford, demonstrated to Henry de la Beche the continuity of the Devon Culm Measures with the Greywacke rocks of the rest of the south-west peninsula. The strata are upright, and continuous along 1 km of shoreline, so represent 3,000 ft of vertical height difference without any break or unconformity. The present-day division between two geological periods does indeed lie within this scruffy little cliff.

BOTTOM Traces of coal in the southern, and so lower and older, end of the Fremington little cliff.

At Tintagel were fossils known to be Greywacke ones. They trekked back along the coast – but the unconformity still failed to appear. No wonder de la Beche had managed to miss it.

The strata were unsatisfactory; and the fossils got worse as well. Annoyingly, some more plants had been found by a Mr Harding at Ilfracombe. This was below a black limestone that they'd earmarked as the bottom of the Culm, so these new fossils had placed themselves high in the Greywacke.

There were three reasons why the big meeting of the British Association, at Bristol in August 1836, didn't end in a ferocious scientific bust-up. Firstly, if de la Beche was wrong, Murchison and Sedgwick weren't actually right. Secondly, anything anybody said was liable to be overturned, as soon as next summer, by the evidence of the fossils. At Torbay and elsewhere, Devon sported four separate limestones asking to be hammered at. And finally, the scientists were determined to distance themselves from the bishops and men of religion by disagreeing with each other in a reasonable and friendly fashion – or, failing that, of plausibly pretending to do so.

So the disappointed reporter from the *Literary Gazette* was forced to report that 'the geological rivals parted as good friends as when they began' – consoling himself with a magnificent pun, that they did so only 'after they hauled each other over the coals'.

Austen's Persuasion

So far in this story, the Geological Society has consisted of wide-ranging geniuses like Murchison and de la Beche. But the subscriptions were paid, and the admiring benches filled, by lesser folk: country gentlemen, parsons with a fancy for fossils, a few fashionable ladies who (like Jane Austen's heroines) enjoyed a walk through scenic landscapes but also enjoyed exercising their brains. Occasionally the inner ring was obliged to let one of these local worthies read a paper before the Society, their fanciful speculations then being politely ignored. After all, they were useful guides to their locales: the best fossil quarries, the sea-cliff with the intelligible strata. But now and then, embarrassingly, they knew more about those fossils than the experts themselves.

One of these locals lived on the crucial Torbay limestones. Robert Austen, of Teignmouth in southeast Devon, was not related to the novelist. But in the best tradition of Jane Austen, he managed to marry a wealthy general's daughter, then gratefully renamed himself as Godwin-Austen and devoted himself to the local geology. On a field trip, he and Adam Sedgwick uncovered some fine coal-measure plants but also overturned (literally) Austen's placing of the Transition Limestone. According to Sedgwick: 'Austen was in raptures at the new aspect of the neighbourhood and while we were knocking out calamites [*Calamites* are giant horsetails] on came a thunderstorm which put us in the condition of half-drowned rats.' Murchison, for one, would not have been so thrilled at being proved wrong.

ABOVE Berry Head, Torbay. Is the sea stack the Mountain Limestone of the Carboniferous, the Transition Limestone discovered by Murchison at the top of the Greywacke, or something else altogether?

BELOW Fossils in the crucial limestone at Sharkham Point, southeast Devon. Left: Coral, but not quite like the coral of the Mountain Limestone. Right: Stromatoporoids – a reef-building organism but one that isn't a coral.

The fossils found by Robert Austen in the limestone cliffs of Torbay seemed similar to the Mountain Limestone brachiopods and corals in John Phillips' book of Yorkshire fossils. John Phillips, the young nephew who accompanied William Smith, had grown up to become Professor of Geology at Oxford and the British authority on Carboniferous fossils – and he agreed with Austen. The Torbay fossils were quite like the Mountain Limestone that lies immediately below the Coal Measures just as the Torbay limestone seemed to lie below the Culm.

But the fossils were only quite like the Mountain Limestone ones. They were presumably just a bit older, or just a bit younger. Younger than the Mountain Limestone would mean the Millstone Grit, or the Coal Measures themselves, both already well known. That left the Torbay limestone as somewhere lower, and older, than the Mountain Limestone.

The next layer down from the Mountain Limestone was the Old Red Sandstone. The Old Red Sandstone is extremely recognisable. Its fossils are few, and peculiar, and consist of fish, and they are not at all like the ones at Torbay. And yet – just below the Torbay limestone, there is indeed a stratum of sandstone. And that sandstone is coloured red.

The map shows Devon, as understood in 1839. After four and a half years of frustration and fieldwork, the clues to solving the Devon dilemma were in place. It was probably Mr Austen from down in Torbay who first set down the answer. But as a provincial, his paper could conveniently be passed over for formal reading by the Geological Society. So it was Murchison who published first, in an article for the *Philosophical Magazine* that started off in lively style with some insulting remarks about Henry de la Beche. (Adam Sedgwick was co-author. The entire controversy was complicated by the observations, sometimes correct and sometimes confusing, of this other great geologist.)

The slates and shales of Devon and Cornwall, hitherto marked on the map as Cambrian Greywacke, belonged to the same system as the Old Red Sandstone. They were, in fact, its underwater equivalent. This laid them immediately

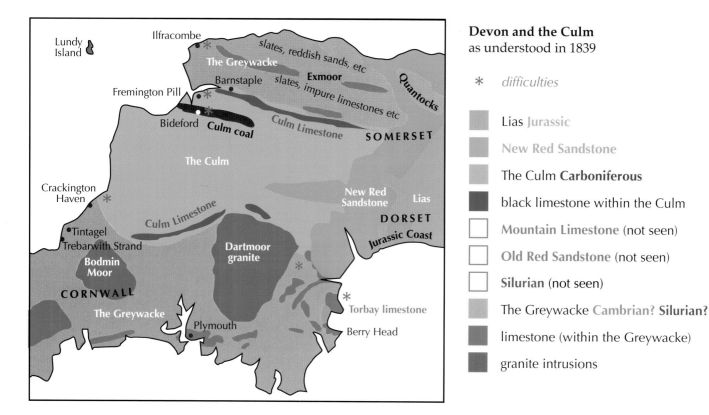

Devon and the Culm
as understood in 1839

* *difficulties*

- Lias Jurassic
- New Red Sandstone
- The Culm **Carboniferous**
- black limestone within the Culm
- Mountain Limestone (not seen)
- Old Red Sandstone (not seen)
- Silurian (not seen)
- The Greywacke Cambrian? Silurian?
- limestone (within the Greywacke)
- granite intrusions

Devon in 1839

The Culm is understood as a trough of younger rocks on top of the Greywacke – the two bands of black limestone are the same stratum appearing at the two sides of the trough.

Difficulties, from top

* Ilfracombe: fossil plants, interpreted by Murchison as upper Silurian
* Fremington: no visible unconformity at the base of the Culm
* Bideford: Coal Measure plant fossils. Are they necessarily of Carboniferous age, or could they be much older?
* Crackington Haven: no visible unconformity at the base of the Culm
* East of Dartmoor: small unconformities, not considered by local amateur Robert Austen as major enough to leap past the missing Mountain Limestone, Old Red Sandstone and Silurian
* Torbay: fossils considered by John Phillips (the expert on Mountain Limestone fossils) as like, but not the same as, those of the Mountain Limestone

below the Carboniferous Limestone, and immediately above the Silurian of Shropshire.

The rocks of Crackington look like the Greywacke. They are layered like Greywacke, they are folded about like Greywacke, and they are the correct shade of grey. But actually they are the Old Red Sandstone.

The thousands of feet of the existing Old Red Sandstone represented a major span of time. The Torbay limestone now inserted into that timespan some fossil corals and shells which were not like either the Silurian ones or the Carboniferous ones, but lay conveniently between them. The geologists had identified, almost by accident, another new geological period. They decided to call it Devonian.

What was called, even at the time, the Great Devonian Controversy, was one of those interesting moments when the recognised way of doing science – in this case, geology – stopped working. The Greywacke system was grey, striped, and recognisable; it lay, stratigraphically speaking, underneath the Old Red Sandstone. However, its fossils were from the Carboniferous, a whole lot earlier.

After six years of thinking about it, the stratigraphy was still confused: Devon is fearsomely folded. The fossils were found to be sound, and fossils are still the first-resort (because cheapest) way of dating rock formations. The tough, grey, layered Greywacke System was a mistake. The Devon bit of it was Old Red Sandstone in disguise.

The Universal Formations are wrong. But they're still jolly useful, and I've used them in this book. Today's theories are, we hope, less wrong. They explain where the Universal Systems had made their mistake. Equally importantly, they explain why, so much of the time and so usefully, the Universal Systems were spot on.

BELOW Looking north to Berry Head. The limestone beds are interrupted by some red-coloured sandstone: a clue picked up by Robert Austen.

Wacke sort of stuff

As the Great Devonian Controversy cooled off, there was still that Culm rock of Crackington Haven sitting in the middle of the Carboniferous. It had dark and light grey stripes, like the Greywacke; like the Greywacke it was fiercely folded. But it contained coal-field fossils. So they looked at it more closely, and realised it wasn't Greywacke at all.

This despite the fact that greywacke was exactly what it was!

The greywackes of Wales, the Lake District, and southern Scotland all represent deep sediments from the edge of the Iapetus Ocean, crushed and thrust up by the Caledonian (or England-Scotland) crunch. The Devon greywacke is the same, but different. It represents sediments from a different bit of sea, thrust up 200 million years later by the Variscan (or Africa) crunch. It was Mr Murchison's bad luck that two similar events formed two similar rocks at two quite different times. As if, two days after you went on holiday, a clumsy digger-driver knocked off the corner of your house; and the day before you came home again, a passing aeroplane dropped a lump of ice through the porch.

The earlier greywackes of Wales, like the later ones of north Devon, are grey, sludgy sediments from the bottom of an ocean. Silt, gravel and sand pile up, a speck at a time, in the stillness and silence at the edge of the continental shelf. One last grain of mud, a falling seashell, or more likely an earthquake, overbalances the slippery heap. Down it goes, over dozens or hundreds of miles, in what is either a very muddy sort of sea-current or else an unusually damp avalanche. The men who make up names have settled on the first, calling it a 'turbidity current'.

LEFT Greywacke, Berwickshire coast. Each thick grey slab represents a single 'turbidity current' or sludgy underwater avalanche, arriving over a few seconds. The worn-away shale between was deposited over the uneventful millennia between the mudslides.

BELOW Where rock layers are as confused and crumpled as most greywacke, geologists are alert for 'way-up' indicators. The picture at Crackington Haven at the starat of the chapter showed slump structures (or load casts), where the tough avalanche layer has sagged into the soft mud beneath it, on the bottom of the bed. This one shows the ripples on top: a reassurance that the understanding of the steeply tilted beds at Marloes Bay – and the deductions about the Silurian and Devonian systems of rocks – haven't been made all upside down.

RIGHT Graded bedding. As the avalanche debris settles, the coarser grit and sand sinks faster. So the bottom of this rock, as seen in the photo, was the bottom when it was first laid down.

The avalanche debris settles into a single slab that's a mix of large particles with smaller ones and fine silt. Like well-graded concrete of sand and aggregate, that mix of different sized ingredients makes a tough rock. Then, for a few years or a few thousand years, fine grey dust drifts to the deep sea bed; plus the occasional graptolite, supposing all this is happening back in graptolite times.

The greywacke, then, is thick, grey, unstructured beds piled one on another. Or else it's those same beds, interleaved with grey sludgy shales. Each thick bed represents a single underwater avalanche, and formed in a few hours or even minutes. The shale between two beds marks thousands of uneventful years between. Now that you know how it happens, you can refer to the stuff as 'turbidite'. Or you can keep on calling it greywacke – and again there's a choice. The word is originally German, and some pronounce that final 'e', to give 'greywacky'.

When Scotland and England arrived from opposite edges of the Iapetus Ocean, the deep ocean turbidites were squeezed up between them, like toothpaste when you tread on the tube. The tough thick sandstone, interleaved with the brittle shale, makes a laminated structure like plywood. Plywood is bendy, and so is greywacke. In Wales and around the Lake District, along the Southern Uplands, and then again in Devon, this is the nature of the greywacke: massive grey beds, but all with lively wiggles and zigzags.

In that first greywacke walk of 1831, the one that started at the Welsh coast and worked eastwards and upwards through the rocks, Alan Sedgwick had a helper, a promising young naturalist who needed a grounding in geology. And he needed it right now, as he was just off on a two-year voyage to the South Atlantic, on a ship called the Beagle. Charles Darwin's first thoughts on evolution were inspired not by the Galapagos finches – they came later – but by a geology book he took with him and read during the first few weeks at sea. Fortunately, the book was a good one (it was *Principles of Geology*, by Charles Lyell).

Seven years later, Darwin was the Geological Society's coral reef man; and he wasn't quite convinced that the stuff at Torbay really was a coral reef. (And it isn't: stromatoporoids are a reef-building sponge.)

Twenty-seven years later, Darwin's theory of evolution gave the first explanation of just why fossils, rather than mere appearance, and more even than position among the rock layers, were the way to decide what your rock was.

Ynys-fach – the name means 'small island' – east of Trefin. The rock is Ordovician slate, with upright cleavage. Waves curve in around both sides of the island to form the two-sided beach, or 'tombolo'.

13. ANCIENT DAYS: PEMBROKESHIRE

> Oh for a horse with wings
> To fly me to Milford Haven!
>
> William Shakespeare, *Cymbeline* Act III Sc 2

By 1835, Murchison was fed up with referring to the rocks of Shropshire as 'Upper Greywacke' when they were actually limestones and sand. He renamed them as the Silurian System, after a tough Welsh tribe conquered by the Romans in the same area. They were led by Caractacus, and, according to the historian Tacitus, could be persuaded to change their aggressive behaviour neither by massacres and atrocities, nor by diplomacy and gifts. Caractacus made his last stand in AD 50, possibly at Caer Caradoc above Church Stretton, the fortified hilltop that bears his name.

But for Murchison, Caer Caradoc was just the third ('Caradocian') subdivision of a resurgent Siluria. Over the next decades, he would display Caractacan stubbornness right across Wales, Scotland and out into Europe. To start with, Siluria was expanding west into mid-Wales, and downwards to annexe most of the Greywacke.

Meanwhile Adam Sedgwick had named the bottom of the so-called Greywacke as the Cambrian system. From Aberystwyth in the east, he was exploring it inland across Wales. The two geologists were converging like continents

Silurian: greywacke, slates, etc

Ordovician: greywacke, slates, volcanics, etc

Cambrian: greywacke, slates, etc

Precambrian

across a shrinking ocean of the unknown. A collision was inevitable, with each claiming most of mid-Wales as part of his personal underground empire.

That controversy was resolved 40 years later (in 1879) by a younger and more orderly geologist. Charles Lapworth used a strategy familiar to parents of young children: 'If you're still squabbling over those middle bits of Wales, then *neither* of you shall have them!' Lapworth was an earnest student of the tiny floating lifeforms called graptolites. The ones of the disputed middle ground were unlike Sedgwick's Cambrian ones, but also unlike Murchison's Silurian ones. They deserved a whole new geological period, the Ordovician, to lie below the Silurian and above the Cambrian. Now that actual time spans can be applied, the Ordovician has 52 million years, and the Cambrian 50 million. Murchison's expansionist Silurian empire has ended as the shortest period of all, a mere 26 million years.

Those heroic-age geologists thought they were placing marker-lines between different rock strata; the fossils were a helpful way to identify those strata. The fossils themselves changed with no sudden jerks or catastrophes – having rejected Noah's Flood, they weren't letting in any Noah's comet crash or Noah's climate change. But they were more modern-minded than they realised. The Ordovician/Silurian boundary, defined (they thought) in terms of rock layers, in fact coincides with the first of Earth's five great mass extinctions.

For a great extinction, there has to be life to be extinguished. The Cambrian marks the lowest lifeforms with hard calcite shells, which is to say, the first fossils. Thus the Cambrian can be counted as the bottom of geology; and the interpolation of the Ordovician completes the set of eleven geological periods.

The Silurian was when Wales, along with England, was at the southern edge of a big southern ocean called Iapetus. Scotland, along with Northern Ireland, was an ocean away at the northern edge. At both those ocean edges there was grey ocean sludge. Eventually the ocean would shrink to nothing at the Caledonian collision, and the sludges would be thrust onto land all folded and bent. And so, as described in the previous chapter, the Silurian features greywacke.

But it also features more ordinary underwater sediments which, after the crash of the continents, would emerge compressed into slates. It features limestones, some of them with those ancient fossils that (eventually) got Roderick Impey Murchison his knighthood.

Under the Silurian lies Charles Lapworth's Ordovician. Scotland was slightly further away across the Iapetus Ocean, so there's some more, somewhat older greywacke, and some more slate. The Ordovician ocean also featured a volcanic island chain. Its rocks, squashed, eroded and gently raised again, now form Snowdonia and the Lake District. But the eruptions also inserted lava and black gabbro among the slates and greywackes. The Ordovician intrusions come in those same greys and blacks, but a lumpy rather than slaty sort of texture.

ABOVE St Ann's Head: Old Red Sandstone, raised and bent about by the Variscan mountain building

BELOW Cobbler's Hole, St Ann's Head

Under the Ordovician lies the Cambrian. The Cambrian is even older greywackes and slates, along with sombre mud-coloured sandstone. And below it all lies the twisted, mysterious Precambrian.

The place to see all this is Pembrokeshire. Pembrokeshire is where Murchison's Silurian runs down to the sea, to inter-mangle with the Ordovician and the Cambrian in a riot of grey, black, grey-black, and every combination between. Massive sandstones, greywackes and slates tilt at 45° to form the triangular sea-stacks of Pembrokeshire.

Keep in mind everything in the previous chapters about sediments and strata, but be prepared to find them tipped about and upside down and further confused by the alterations caused by crushing. Some of what you see as sedimentary beds is actually slaty cleavage, the rocks separating into layers as a result of compression. You must then remember all the possibilities of igneous intrusions. None of those intrusions will be colour-coded purple or red: they will be black and grey, the same as the sediments. And anything, anywhere, can get thrown out by fault movements.

It's like Grand Theft Auto; as you get better at it, you move up a level. But in geology as you get better, it's time to move on down …

The Ria Grand

The complicated branching inlet of Milford Haven is the equivalent, in water terms, of Pembrokeshire's land peninsula itself sticking out into the Irish Sea. The Pembrokeshire Coast Path from the Angle Peninsula treks east, south, north, west, and southeast over two long days – to arrive at Dale Point, two miles away across the water. In the quotation from Shakespeare's *Cymbeline* which started this chapter, Imogen wished for a winged horse to fly to Milford Haven. That wish may be shared by the coast path walkers as they make their way past the mighty oil refineries.

The inlet wasn't carved to below sea level by any glacier – it's not a fjord. It is not any sort of bay either, because it was not carved by sea erosion along faultlines or less hard rock layers. Seen on the map, the haven has the same multi-pointed shape as reservoirs like Kielder Water in Northumberland. And it's formed the same way, a complex valley carved out by River Cleddau and its tributaries, then flooded back by the sea. In Chapter One, raised beaches were evidence of land that has risen since the Ice Age, due to the removal of the overlying ice. Here is the opposite, land that did not

rise and accordingly has been flooded by the rising sea as the world's icecaps melted away. As the ice covered north Pembrokeshire 18,500 years ago, the sea was about 40 m lower, and you could walk from Tenby to Barnstaple across the Bristol Channel.

Inlets like this - river valleys flooded by a rising sea - are called rias. It's a Galician word from Northern Spain that's the female form of rio, a river. (Fjord, as the spelling suggests, is Norwegian.)

In 1588 a chain was stretched across the inlet to keep out the Spanish Armada. Today, the entrance feeds in a daily line of oil tankers to the refineries, which were built in the 1960s. Their cargoes, sludged-down fossil bacteria, could be said to add the Jurassic to the six geological periods already shown along the Pembrokeshire coast.

At Solva, southeast of St David's, you don't need a helicopter view to see that the harbour inlet is a drowned river valley, or ria.

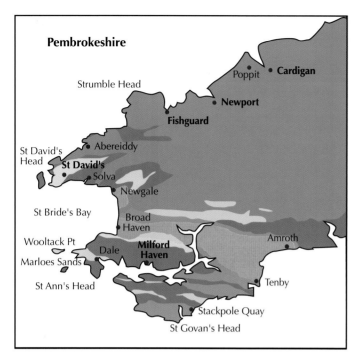

Pembrokeshire

Carboniferous: Coal Measures

Carboniferous: Mountain Limestone

Devonian: Old Red Sandstone

Silurian: greywacke, slates, etc

Ordovician: greywacke, slates, volcanics, etc

Cambrian: greywacke, slates, etc

Precambrian: volcanic

Mountain Limestone

Coastal geology is all about understanding what you see. But what you see is so often lovely to look at. Pembrokeshire starts with 30 miles of Mountain Limestone, all around St Govan's Head. The Green Bridge of Wales is possibly the UK's most spectacular sea arch – supposing they let you in through the tank warfare ranges (as they do most weekends – see www.pembrokeshireranges.com). All along this first coast, keep reminding yourself it's Carboniferous Limestone, tilted by the Variscan mountain-building. Its fossils show it was a coral reef; it dissolves in seawater to make these wonderful shapes. I understand some of this stuff, I really do. Hold hard onto that thought among the black mangled rocks and sea stacks to come …

We're heading anticlockwise – I hesitate to say northwards on a coast so doubled back and twisty – in which twistiness it is only imitating its own rockforms, of course. As we progress the sea will be on our left (if also behind the headland on our right, not to mention somewhere else away ahead beyond St David's). At Freshwater West, you move down in time, onto the Old Red Sandstone that forms the entrance cliffs on either side of Milford Haven.

Old Red Sandstone

The dale at Dale is a good 400 metres wide, running right across the peninsula to link the Irish Sea coast with Milford Haven. It was formed by a fair sized river running down out of thin air 40 m above the seawater of Westdale Bay. This makes perfect sense if you imagine the Irish Sea filled with one enormous glacier, and that fair sized river running down off the ice. The valley conveniently cuts

out 8 km of coastline. I ignored that short cut for reasons of simple spelling. I'd already photographed the stones of St Bees Head and St Abb's Head. This made it impossible for me to bypass St Ann's one.

Old Red Sandstone is a respectable rock. Formed from rubble washed out of the Caledonian mountain range, it arrived just too late to be upheaved by the Caledonian mountain building. So it arranges itself tidily, in nice flat layers.

But here in west Wales, there's been another mountain crunch. The ORS of west Wales has been shoved around and crumpled, not by the Caledonian, but by the more recent Variscan. At St Ann's Head, between the old lighthouse and the current one, a scruffy path between two fences leads out to Cobbler's Hole, and a spectacular S-fold in the massive sandstone.

Of the triangular sea stacks of Pembrokeshire, about 30 stick out of the sand of Marloes Bay. The triangles, like the spines of some decorative aquarium fish, come in two contrasted shades: pale grey, and pink. For Marloes is where the Old Red Sandstone descends into the sea sediments of the Silurian. As described in the previous chapter, the change is 'conformable'. The sea sank to reveal the grey sludgy sediments, and then the sands of the red desert blew in over the exposed sea bed; there is no time gap. And we (and Roderick Impey Murchison) can be satisfied there's no call for another geological period below the Old Red and above the Silurian.

The red sinks back into the grey at the back of the bay; and it does so three times over. Which is just as well. Coming from the south over the Old Red of St Ann's Head, the first junction you come to is buried under rubble. That rubble is a mix of red rocks and grey, but that isn't at all helpful. At the other end of the bay, the red returns. The junction is clear: it is clearly a fault. The red has slid down across the grey on a line cutting the strata diagonally. Third time lucky: behind Gateholm Island, the Devonian consents to lie conformably against the underlying Silurian in the way the geology books require.

Around the Marloes peninsula towards Wooltack Point, the pale underwater sandstones alternate with basalt lava-flows. The lava is harder, and forms the headlands; the sandstone between makes the bays.

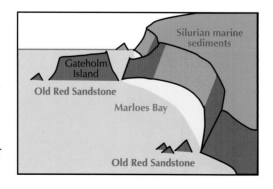

ABOVE The second junction of Silurian with the Old Red Sandstone, at the west end of Marloes Sands. We were hoping for a conformable junction; instead, it's an unusually clear fault plane. To left, ORS and grey Silurian interleaved. The conformable junction to the west of here is shown on page 195.

LEFT Silurian fossils at Marloes Sands: some very small shells, and some even smaller crops of coral (one coral directly above the coin, another at 7 o' clock).

BELOW Fossil worm burrows, Marloes Sands

This is convenient when it comes to working out which is which. It's even more convenient if you're a seal. The lava-and-sand arrangement leads to a succession of inaccessible bays, just right for leaving your pup while you go and eat fish. At the Ranger Hut – the rangers, like the rest of us these days, do their ranging indoors gazing into a computer – I wondered why there were so many dead pups lying around like abandoned bags of cement. I asked sadly, but the rangers just laughed. The pups were lying around, but not dead. They were snoozing after a feed of fishy-tasting milk.

St Bride's Bay

Among Pembrokeshire's bays St Bride's is the big one, as wide as half the county. The sketch map suggests that the sea has moved in along the softer coal measures. Except that in Pembrokeshire, even the coal measures aren't particularly soft.

In the previous chapter, Henry de la Beche failed to spot these coal measures, two geological periods younger than the greywackes and slates to either side. It's more obvious if you live in the twenty-first century and you've already peeped at the answers in the back of the book. But also, de la Beche may have been exploring not Newgale, with its almost-cheerful yellowish sandstone, but Druidston Haven. Coal Measures are crumbly, and hardly ever form spectacular cliffs. As usual, though, Pembrokeshire is the exception. At Druidston Haven, the Coal Measures rise high and

overhanging above the beach. They are also black, so that they match the much older rocks alongside.

Underneath the Old Red Sandstone, the Silurian strata hereabouts were underwater slates and greywacke. More greywacke and slate, Ordovician in age, form the northern shores of Pembrokeshire. At Druidston Haven, it's the slates we see.

By leaving out from this book the Highlands of Scotland (hardly prime sandcastle country), I could have stopped at the Old Red Sandstone and left out the three earliest ages. Well, I could if not for Pembrokeshire. By leaving out Scotland's mountains (the Moine Schist, the ancient gneiss) I could have skipped past the third of the three main rock types. I could have written Chapter Two on the sediments, Chapter Three on the volcanic rocks, and then ignored the messy metamorphic ones. But Pembrokeshire cannot be ignored.

The metamorphics are the squashed rocks, the ones created out of earlier ones by heat and compression underground. The simplest of such rocks is slate. Flat minerals within the rock line up at right angles to the squashing direction. This makes it easy for such slate rocks to split apart along those planes. In smooth sorts of stone with the correct mica minerals, the resulting slates can be split thin and strong to roof your house. Out on the sea cliff, the flat cleavage divisions look like bedding planes – but they aren't, and they run in a completely different direction.

Across Wales and north Cornwall, the Variscan has achieved uplift and also, at the same time, cleavage. The Variscan is indeed the Wonderbra of UK earth movements.

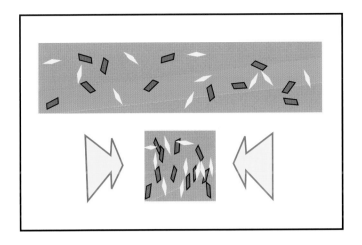

LEFT Renny Slip to Skomer Island. The headland, as well as Skomer Island beyond, are basalt lava. The nearer bay is eroded out of the softer sandstone.

BOTTOM LEFT A seal pup learns to swim in an inaccessible stony bay south of Wooltack Point.

ABOVE The nature of slate. Clay minerals are silica, which is needle shaped, and mica, which comes in little flat plates. Movement within the rock, as it squashes down, will line these crystals ever more at right angles to the direction of squash.

BELOW Ordovician slates at Pwlldawnau, north Pembrokeshire

Ordovician

'Silures istos tenarciores esse aestimasne?' So you reckon those Silurians are pretty tough? You haven't met the Ordovicians, mate. And the rocks named after them demand a tough-minded geologist. The scattered lumps of black gabbro are like the plum stones in a badly overcooked Christmas pudding. You can knock bits off and send them to be sliced in the lab to look at the crystals. Or there's the sea itself, which will provide a polished set of volcanic pebbles – and let you guess just which cliff each one was washed out from.

Then there's the greywacke. The thick sandstone beds are avalanche debris, and don't hold any fossils. There might be slump structures to let you work out which is the underneath side of the bed, just supposing you don't stupidly mistake them for the ripple marks on the top. Any bedding you see, those are slates and it's actually cleavage; while any cleavage could just be greywacke after all, and actually bedding.

After a day and a half of Ordovician clifftops, I headed up beside the crumbling slate works at Porthgain. The clifftop is scratched across by the slate tramway, pimpled with small slate piles, with the added acne of the occasional fallen shed. And I peered down into the great, empty slate mine at Abereiddy.

In a triumph of public relations, the derelict slate quarry at Abereiddy has been re-labelled as the Blue Lagoon. Somebody at the Tourist Board deserves a pay rise, I'd say.

Far below, a photographer had been seduced by the creative naming, and was scurrying around with a tripod. That photographer wasn't disappointed. The storm of the day before was still sending breakers rolling in from northwards, with foam bright among the black sea stacks. Early sunlight slanted across the quarry hole, to extinguish itself against the slate-black walls. The sea water was living up to the tourist brochure by being bright blue. And after picking through the debris for 20 minutes, I found an Ordovician fossil.

It wasn't a spectacular spiral ammonite – that's Jurassic – nor was it the patterned rays of a Carboniferous coral. Ordovician fossils aren't like that. It was an ocean floater called a graptolite, a scratchy mark a good centimetre in length. It would have been fascinating to spend all morning, perhaps with a hammer, and uncover the even longer and two-branched graptolite *Didymograptus*.

On the other hand it was more fascinating to carry on along the coast, enjoying the sunshine and the blackberries and the changing sculpture park of sea stacks. Down on Abermawr beach was a seal, with her pup. The seal was urging the pup towards the sea because there was a dog on the beach. On land seals move in humping wriggles, like a person in a sleeping bag. The pup chose bad routes between the boulders and got stuck, and the dog came over to see what the splashing was about, and the mother lured it away in a different direction. 'As good as a soap opera,' we said on the clifftop, as we clicked our cameras at the harrowing scene.

Cambrian

Below the greywackes and slates of the Silurian, the slates and greywackes of the Ordovician. Below the Ordovician, guess what? At the north end of Newgale Sands, the Cambrian of Wales (which is the authentic Cambrian Cambrian) starts off with some greywackes. The telephone book bends more easily than a slab of timber. And the greywacke, with its tough sandstone beds separated by layers of shale, bends at Pwll March into curves like Sydney Opera House.

At the start of the Cambrian, Nature invented the seashell. And suddenly, over a time that may have been as momentary as 5 or 10 million years, 100 different basic body plans came into existence. Since then 70 of the basic body-plans, or 'phyla', have become extinct. The 30 that survive include the Echinoderms (crinoids, sea urchins etc), the Arthropods (beetles, insects, lobsters, comprising 80 per cent of current animal species), and the Chordates, creatures with backbones, which includes us. Since that Cambrian Explosion, there have been new species and new families of species, but no new phyla at all.

Of this abundance of new life, what shows in south Wales is a small shellfish called *Lingula*, a bit smaller than a 5p piece. Of the 30 phyla, *Lingula* belongs to the Brachiopods, which have been shellfish from the start right up until the present day, when a couple of dozen species inhabit the Arctic seas. The Brachiopods are an entirely separate sea shell in a different phylum. As two-sided animals with an anus, *Lingula* are more related to human beings than they are to limpets, clams and ammonites in the Mollusc phylum.

TOP Graptolites, after 400 million years in the dark, now sunning themselves above the Blue Lagoon

ABOVE The right-hand rock has strata running in diagonally, then cut off at the junction plane. This looks like an angular unconformity, one set of sediments laid across the eroded ends of an earlier set (like Hutton's Unconformity see Introduction). But what appear to be bedding planes are in fact slaty cleavage. The black sandstone, on the right, contains the right mud minerals to realign themselves slatewise. The underground volcanic gabbro on the left was tough enough to resist the squeeze; and anyway doesn't contain those mica and silica minerals. The junction seen here is, accordingly, the edge of the gabbro intrusion.

Here in the Cambrian of Cambria, *Lingula* inhabits a grey-and-brown striped sandstone and mudstone known as the Lingula flags. Flags here mean flagstones: the yellow sandstone beds are a convenient thickness for paving your old Welsh kitchen. The stripy beds have been raised on edge by the usual earth movements, and carved by the sea into sea caves and the occasional triangular sea stack.

Westwards along the Solva coast, the Cambrian offers other sorts of underwater sandstone in blackish yellow and bruised-plum red, all of it compressed into slates. And at the very bottom of the Cambrian, there's a layer of embedded pebbles, a conglomerate. This junction is an unconformity. The conglomerate is the bleak beach of 550 million years ago, as the Iapetus Ocean carved into sea cliffs of volcanic rocks. Those rocks were ancient even then. They date from before the Cambrian, before the first shelly fossils, and so, in a sense, before the beginning of geology.

Precambrian

The Precambrian isn't a geological period. It's seven times as long as the 11 geological periods added together, and constitutes all the time before hard-shelled, handily fossil-making life. Since the 1960s it's been possible to date with actual millions of years, using radioactive elements within some igneous rocks. This makes the Precambrian slightly less mystifying. But the Precambrian rocks have been knocking around an awful long time, mangled by a whole sequence of continental collisions, and buried deep enough to get cooked and squeezed even in the quiet times.

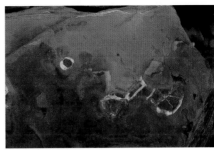

TOP The black is mudstone: the pale grey is coarser sandstone. The sand fills a channel in the mud carved out at right angles to the present rock surface. Traces of slaty cleavage, slightly left of vertical, are in the mudstone but not the sandstone.

MIDDLE Shells in the Lingula Flags. The small brachiopods (upper left, 1cm across) aren't actually *Lingula*, which are oval in shape – they resemble fingernail extensions.

BOTTOM Lingula Flags at Trwynhwrddyn, Whitesands Bay west of St David's

Ignoring as always the mountainous bits of Scotland, these deep rocks appear at less than a handful of spots. One of them, needless to say, is within perplexing Pembrokeshire. Around Pen Dal-aderyn, southwest of St David's, the rocks are of upper Precambrian, or only moderately ancient, age. They are mangled, crushed, volcanic lavas and ash, mixed in with a little mangled granite. The Precambrian coast takes a couple of hours to walk around. As I approached from St Justinian, cloud crossed the sky to veil the inscrutable, amorphous rocks in a suitable dimness. As I emerged again west of Porth Clais, sunlight would twinkle again on the, by comparison, cheerful black sandstones of the Cambrian.

TOP Lumpy Precambrian volcanic rocks at Pen Dal-aderyn, southwest of St David's

MIDDLE Multi-coloured Precambrian volcanic pebbles, Porthlysgl Bay. The three central stones are tuffs, of pumice chunks and volcanic ash. The pinkish one shows pumice specks flattened by overlying ash while still semi-melted.

BOTTOM Precambrian gneiss at Mhealasta, Isle of Lewis

FAR RIGHT Fishguard at night

The lift has eleven buttons: eleven storeys of life and variety. Sixth floor, there's a blue tropical swimming pool. Next floor above, useful building materials; then the fish restaurant and the rooftop garden. At the very bottom, there's the lift-button marked basement. Down in the basement there are windowless rooms of dusty storage, and builders' rubble, and an ancient, dangerous boiler.

At the very bottom of the continental rocks, below the sandstone and the basalt and the chalk, the granite and the greywacke, is the Precambrian basement. It's a mangled and metamorphosed rock, classed as schist or as gneiss. It's crystalline, but not in the sparkly style of volcanic granite or gabbro. It's striped, but not with the simplicity of sandstones; the schisty stripes are bent about and tangled. Its age is in not millions but billions of years, and it floats on the black semi-rigid matter of the Earth's next layer in, the Mantle.

The basement rocks emerge at the hearts of the ancient continents: in Canada around Hudson's Bay, and central western Australia. Here in the UK, the Lewisian Gneiss of the north-west Highlands and the Western Isles is an ancient continental middle. The Lizard peninsula shows a strip of schist, the ancient basement, before descending into even weirder sorts of stone. And on the south coast of St Bride's Bay there's one speck of what may be Precambrian basement, the Dutch Gin Schist. Does it call for a geologist with a very strong head? It may just be named for its location, dangling above Dutch Gin Bay, 300 metres northeast of Mill Haven.

Strumble crumbles

Strumble Head's Ordovician-age gabbro and pillow lavas are the blackest part of all these dark cliffs. Or maybe it was just the gale from the north, speckling the grey sea with white-caps, and at the cliff foot raising towers of spray that drifted across the path to mix salt water with the rain on my face. The rocks were shapeless in the grey light. The wave spray crashed over, then drained in white strings between the lava pillows.

It was natural to see the waves as ineffectual, the rocks as standing firm. Nothing could be more wrong. These Ordovician rocks, formed 450 million years ago at the base of a volcanic island, raised and rearranged 100 million years later by the Scotland collision – preserved under younger rocks as the future UK drifted northwards across the equator, nudged suddenly sideways by the opening of the lively little Atlantic – these aged and travelled rocks are being dismantled, by the water, in a geological eyeblink. A slowly-rising ocean eliminates the business with the wave-cut platform; so global warming could further reduce the Strumble's crumble time to half an eyeblink.

Dusk crept out from the rock shadows to take over the sky. Where the cloud tore apart, the rifts were bright-edged with moonlight, but their centres were black night sky and stars. Strumble lighthouse sent ghostly patches zooming across the cloud.

The path headed between wind-noisy thorns, then down through a lifeless housing estate, to cross a caged-in way above four lanes of concrete heading for the Rosslaire Ferry. An orange sign told me and the empty concrete what the wind speed was out there. A tarred path, between sea and clifftop, runs around the edge of Fishguard. Below me, wave foam shot up into the edge of the streetlights. Across the dark harbour, the lit fronts of old harbour cottages stood in a homely row.

14. ROCK BOTTOM

Steeple Rock, a serpentine stack at Kynance Cove. (The pink rock sticking out of the sand is a surprising fragment of gneiss.)

14. ROCK BOTTOM

Out to sea, Ailsa Craig sticks up like an onion – it is in fact a named onion variety, favoured by those who grow monster vegetables for the show table. It's also a tomato, so that Ailsa Craig is unique among islands in having two separate varieties of veg named after it. Ailsa is the unlikely island in the background of televised golf from Troon. It is, like most upstanding islands, the plug of a former volcano; this stump of microgranite was the lava feed to a crater thousands of metres above.

That volcano popped up a mere 50 million years ago. The mainland rocks underfoot are much older. A steep little path leads down to a bay where the main road above is quickly forgotten. The sea washes around black boulders, and a fault-line has created a crack of a cave just above the high-tide line. Inside, the cave widens out, with a floor of gravel and the droppings of either birds or bats. Something with big wings flaps in the darkness overhead.

Faultline caves usually narrow at once to rubble. This one has probably been plucked out by the sea, at a time when the sea level was higher. So this cave could be inhabited in only moderate discomfort. And it was lived in, at least in legend. A broadsheet of the seventeenth century reveals it as the bouldery home of Sawney Bean and his family of cannibal bandits.

The Ayrshire coast road runs along a line of raised beaches. But above this bay it's a shelf carved out of the steep hillside. Here Sawney and his family, all conceived via incest, would catch travellers and pickle their limbs in seawater. Sawney slipped up, and one of his captives escaped. Even so, a search of the shoreline failed to uncover the cannibal bandits, as the high tide happened to be flowing into the front of their cave.

ABOVE Sawney Bean's Cave, at Balcreuchan Point, Ayrshire. It has developed along a fault, and was probably sea-plucked at a time when sea level was a few metres higher than today. The cave is in basalt lava, but the foreshore rocks below are blacker, and stranger.

RIGHT Pillow lava south of Sawney Bean's bay

Eventually, King James himself came with sleuth dogs and uncovered the cave, its miserable cannibal inhabitants, and the joints of human ham hanging from the ceiling. The menfolk were hanged. The women's crime was worse, as it is the duty of men to eat what their women set before them; they were burned to death.

Cannibalism, banditry, incest, and an uncomfortable life in a cave. But the dark, rounded boulders in front – where Sawney and his family spread their evil feasts – are every bit as unnatural and bizarre. Some are mottled blackish-green. Others are streaked with white, like the lard-rich limbs of some plump murdered traveller.

The walls of Sawney's cave are black basalt lava, shattered by the faultline that formed it. Nearby coves show basalt in big rounded blobs of pillow-lava, squeezed out deep under water. Around the lava are odd rocks of several other sorts. Three miles up the coast is gabbro, produced in a deeply-buried magma chamber, but here with large white and black crystal patterns. On the hilltop above is trondhjemite: a pink granite, but without granite's black specks of mica. In front of the cave, the streaky white rock is tangled-together veins of carbonate and silica. The greeny-black rock is mostly composed of the basalt-family mineral called olivine. The rocks as a whole, just like the Bean family which inhabited them, are outside the civilised scheme of things. You can travel 500 miles through Britain and see nothing like them.

But then at the 501st mile, you arrive at Cornwall's Lizard.

Snake stones of the Lizard

Take water and oil, and beat them together so that they stay suspended in each other. The result – butter – is an emulsion. Do the same thing with a gas and a solid (or a liquid), and it's an aerosol. Kynance Bay is an aerosol, a beaten-together mix of sea breezes and geology.

The space above the beach, where you'd expect air, is about 50 per cent sea stack. Head in between the stacks and you're on a different beach, one pointing out in the opposite direction. That beach, naturally, is interrupted by a sea stack. Now, examine the rock walls behind the stacks. That rock is 20 per cent empty air. Caves run through the cliff, caves floored with beach sand – and you're back on the beach where you began.

The sea stacks themselves are black, black with green streaks, black and maroon red, with the occasional yellow bits. They look unearthly, and they are. They are also slippery to scramble about on.

Elsewhere on the Lizard there is grey schist, with zig-zag stripes and the glitter of mica crystals. There is gneiss, tough and rounded, some of it pink, some yellowish dotted with grey. And there is gabbro, crystalline black with white speckles. Quite orthodox rocks, at least so long as we're talking about the bent and ancient crags of the Scottish Highlands, or the sudden eruptions of the Isle of Skye. But what is all this doing at the extreme other end of Britain?

The gneiss and the pinker sort of schist are, at least, British.

The Lizard

Gweek
Helford River
Giant's Rock *
Porthleven
Helston
Porthallow
Mullion
Coverack
Mullion Cove
Kennack Sands
Cadgwith
Kynance Cove
Church Cove
Lizard
Lizard Point
Man of War Rocks *
Polbream Cove

TOP RIGHT Man of War gneiss pebble at Polpeor Cove

RIGHT Mica schist seacliffs of the southwestern Lizard. The sparse, heathery foreground is growing on serpentinite; the schists give a richer landscape of pastoral grass.

country rock: marine Devonian slates
····· boundary fault
Precambrian schist
Precambrian Kennack Gneiss
gabbro
serpentine
Man of War Gneiss

222

We know more about the distant stars than the earth a mere dozen miles below our feet. Stars are hot gas, which can be analysed with the spectroscope. You find out about the Earth from earthquake shocks passing through it – as an ultrasound scan can see an unborn baby. To actually examine the rocks from 20 miles down would take a rather large shovel.

But a medium-sized continent does the trick. The deep rocks of Girvan were squeezed to the surface by the collision of England and Scotland, at the end of the Silurian Period. The deep rocks of the Lizard came up as Europe collided with Africa in the Variscan crunch, at the time when the super-continent Pangaea assembled itself towards the end of the Carboniferous.

If you could draw up the mythic Kraken from the deep, it would explode due to decompression. It would then shrivel under the unshielded sun. The same goes for these deep-down Mantle rocks. They don't like it up here. Up here has two corrosive chemicals: air, and water. If you could reach down a volcano and get it, a fresh sample of the Mantle would be green crystals of olivine. But as the deep rock emerges over thousands of years, it emerges decomposing.

The rock was called serpentine because of its snakelike look and texture. Today, 'serpentine' is the group of greenish, slippery-soapy minerals, and the rock made of them is 'serpentinite'. But the State of California, whose state rock is serpentine, confuses the two terms, so the rest of us can as

They are the Precambrian basement rocks of the continental crust, so they have been knocking around the world since the world congealed out of red-hot rock, and the scum of silicates floated to the surface.

The gabbro and Girvan's pillow lavas are not continental rocks at all, but oceanic crust. Continental crust, rich in silicates, is pale coloured and relatively light. The ocean crust is black and heavy. It emerges at mid-ocean ridges as these basalt pillow lavas, dolerite, and gabbro. Both continental and oceanic crust float on the Mantle, which is even heavier, even blacker, and seethes around in a neither-solid-nor-liquid state, driven by the Earth's internal heat.

well if we want. The group of strange stones which includes serpentine, pillow lavas, and a deep-sea sort of silica lump called chert, is collectively referred to as 'ophiolite', which is a Greek (rather than Latin) way of saying snake-like.

This still leaves the yellowish gneiss among the pebbles of the Lizard's Polpeor Cove. The Variscan mountain belt is well to the south of the UK – the Pyrenees, and the Atlas mountains of north Africa. However, as the continental masses shuffled into position, England did receive a bash from the corner of Brittany – at that time attached to the southern lands called Gondwana. A smear of Gondwanan gneiss forms the Man o' War Rocks, just off Lizard Head. Part of the foreshore is similarly a snatch of ancient Gondwana, but you'd need to be a climber or else a shipwrecked sailor to find out for sure. So just head around the corner to Polpeor Cove, where boulders and pebbles of Man of War gneiss lie half-buried among the mica schists and the sand.

The stones of Kynance are from 40 km underground. But here, and only here on the UK mainland, you can spread your beach towel over a fragment of ancient Africa.

Seeing the beach

The sea-shore is a sort of neutral ground, a most advantageous point from which to contemplate this world ... Creeping along the endless beach amid the sun-squall and the foam, it occurs to us that we, too, are the product of sea-slime.

Henry David Thoreau
'The Sea and the Desert' in *Cape Cod* 1865

Lying in a green nylon bag, 100 m above the North Sea as the stars wheel overhead, it's easier to think big about the ocean.

Other planets – plus some of their moons – have volcanoes. None has moving tectonic plates, as Earth does. They don't have mid-ocean ridges, or Himalayan mountain ranges where moving continents collide. They don't have subduction zones where surface rocks are dragged back into the planet. One third of Earth's surface water is water of crystallisation, locked into the minerals of the sedimentary rocks. As ocean crust is dragged into Earth's interior, water goes with it, and some of it emerges again out of volcanoes.

Sometimes the slow dance of the continents creates several new mid-ocean ridges simultaneously. This happened in the Cretaceous Period, the time of the chalk, to raise the sea level by hundreds of metres. Even in the current short century, human-induced melting of the icecaps may raise the ocean by several metres. The Ice Age, a few tens of thousands of years ago, lowered the sea by 80 m. But for as far back as we can see – and that's about 4.2 billion of Earth's 4.5 billion year history – there has been ocean, and there has also been dry land.

Other sunlike stars have planets: the Hubble telescope can even see some of them. But how many of those planets are Earthlike and might sustain life? With our outsized moon, our moving tectonic plates, our rocky continents rising out

LEFT Man o' War Rocks, a fragment of ancient Gondwanaland

BELOW UK's southernmost cafe, above Polpeor Cove

FOLLOWING PAGE Sunset below Cowbar Nab, Cleveland. An evening surfer wanders across the wave-cut platform.

of blue ocean: is the Earth a planetary fluke – or are we much, much more surprising than that?

Our trick, which we think is pretty special, is running imaginary simulations in our minds to plan out what might happen. We don't just think: we think about ourselves. To a dinosaur, this would be a trivial trick, not to be compared with running through a swamp at 45 miles per hour.

We laugh at the dinosaurs who left their snouts sticking out of the Lias layers around Whitby harbour and all along the Dorset coast for going extinct in such a silly way. The dinosaurs persisted for 200 million years. Humans have been around for less than one hundredth as long. And as civilised beings monkeying around with our entire planet and its atmosphere, we've only been at it for two short centuries. Some have solved the problem of why we've met no other thinking beings in the Universe by supposing that the lifetime of an advanced mechanical civilisation is a geological instant, a few hundred years at most. We don't see anybody because all the others are already extinct.

Is every intelligent creature doomed to destroy its own world quite so quickly? After all, it's quite conceivable that even we humans might survive the mess we're currently making of our world.

On the other hand, life may be as common as six-times Lottery winners. We could even be the only ones like us in all the whole, wide Universe.

Be a shame to spoil it all at this point, wouldn't it?

NOTES

Introduction

The Ordnance Survey has calculated the **length of the coastline** of Great Britain (the one big island only, at mean high water mark) at 11,072.76 miles. On the other hand, the CIA gives the UK coastline (including, this time, the 400-odd miles of Northern Ireland) as 7,732 miles. The fact is that our coast doesn't just have wiggles, but wiggles on the wiggles. How much coast there is depends entirely on how closely you look.

One mathematician has estimated this 'fractal' quality of our coast at 60 per cent. Each time you look twice as closely, any previous ten miles of the coast now measures out at 16. The OS figure was measured on their large-scale mapping of 1:10,000. If the 60 per cent factor is correct, we can calculate how long the coastline would be if you walked all the way around it with a tape measure. The result is a rather astonishing 5,708,749 miles. If that's not enough, you could get a far larger figure by measuring through a microscope.

The OS claims to have measured this unmeasurable coastline to an accuracy of 10 m. This is so dubious that in this book I've unpatriotically gone for the CIA's figure.

John Playfair quote (**abyss of time**): *Transactions of the Royal Society of Edinburgh*, vol V 1805

Siccar Point on TV: the series 'Men of Rock' presented by Prof. Iain Stewart aired on BBC Two in Scotland only during winter 2010/11

Chapter 2. Understanding Sand

Spellings of 'mollusc': Gary Rosenberg, in 1996 issue of *American Conchologist*. His earliest reference for 'mollusc' is the great Scottish geologist, Charles Lyell.

Chapter 4. Hot Rocks

Man with gun: Raymond Chandler, *The Simple Art of Murder* (Introduction) 1951. To the list of gunman-inserted works we might add *Sandstone and Sea Stacks*, its gentle coastal ramblings interrupted here and there with a lumpy, sharp-edged geological fact.

Chapter 5. How to Make Mountains

The **two sorts of coral** current in the Carboniferous (tabulate, rugose) both became extinct in the end-Permian disaster. The kind we have now is a third sort. So the assumption that Carboniferous coral implies a tropical climate is a weak one – rugose corals might just have been more rugged. Indeed, there was coral here in the Jurassic, when the UK was no further south than Mallorca is now, but the climate of the world as a whole was warmer.

Cold War technology provided yet another confirmation of plate tectonics. In the 1950s, seismometers were set up to detect Russian nuclear explosions and monitor the Test Ban Treaty. They were sensitive enough to detect not only the position but also the depth of earth disturbances. Thus they revealed that earthquakes got deeper as you moved in from the continental edge, and were being generated, precisely, along the descending top surfaces of subducted plates.

Chapter 8. Jurassic Coast

Jurassic climate cycle: 'Modelling Late Jurassic Milancovitch climate variations' by PJ Valdes, BW Sellwood & GD Price Geological Society, London, Special Publications; 1995; v. 85; p. 115-132. Other geologists have linked the Lias layers with a different Milancovitch cycle, the 38,000 year one.

Chapter 9. Yorkshire Rock

The **Bahamas Banks** comprise the Great Bank and the Little Bank, but not the Bahamas Central Bank, which is a tax haven. Its unique $3 bill does carry an image of one of the oolite beaches. Limestone sand is unusual. Warm water can contain less dissolved calcite; the Caribbean is warm; and around the Bahamas, calcite is precipitating out of the warm seas rather than being dissolved into them. Apart from not being a tax haven, Jurassic Dorset was the same.

Chapter 13. Ancient Days: Pembrokeshire

Precambrian basement rocks are also found on Anglesey and the Lleyn Peninsular, at the northwest corner of Wales.

Cymbeline, Shakespeare's Welsh play – with its kidnapped princes living as simple hunter-gatherers, its villain Cloten dressed up in the hero's clothes and then managing to get his head cut off – could have been written after too much looking at Pembrokeshire's contradictory rocks.

By a small coincidence, I was examining Pembrokeshire in the autumn of 2010 – the 400th anniversary of *Cymbeline*'s first performance. By an even smaller one, the younger of the two lost princelings in the play, Arviragus, is alternatively named Caractacus: the man who was to become the battle leader of the Silurians. If his cave was just a few miles north of Milford Haven, then it was in appropriate rocks of the Silurian system.

ROCK KNOWLEDGE

Websites

www.discoveringfossils.co.uk
By Roy Shepherd. A useful fossilhunter's guide to about a dozen beaches, mostly on the south coast.

www.ukfossils.co.uk
UK Fossils Network. Fossil hunting info for dozens of locations. Most are coastal – known fossil places inland get used up, but cliff collapses replenish the shore.

http://www.staithes-town.info/geology.htm
Staithes Town Web (author unknown) Lucid and useful.

www.soton.ac.uk/~imw/index.htm
Geology of the Wessex Coast by Ian West. Very detailed, full of pictures, and with some of the topics right at the high tide line of current research. 'Wessex' is south coast, Sidmouth to Brighton.

www.jurassiccoast.com
A straightforward introduction to the Devon/Dorset Jurassic Coast.

www.englishrivierageopark.org.uk Torbay geopark

www.geolsoc.org.uk/gsl/education/page2544.html
Gower Peninsula in considerable detail.

website called 'Roadside Geology of Wales'
By Jim Talbot and John Cosgrove. Most of the roadside geology turns out to be seaside. A lucid and useful site, not technical, and (unlike some) most of what's described is clearly visible and interesting. Beware a slight tendency to write 'east' by mistake for 'west'.

www.pol.ac.uk/ntslf/tidalp.html – tides for the next 30 days: note that times are in GMT not British Summer Time.

BOOKS

Chevalier, Tracy *Remarkable Creatures* (HarperCollins 2009)
A novel about Mary Anning and Lyme Regis. It's hard to resist Chevalier's project of 'making fossils sexy' – achieved not just by giving Anning a fictional love interest.

Cutler, Alan *The Seashell on the Mountaintop: a story of science, sainthood, and the humble genius who discovered a new history of the Earth* (Dutton Books, 2004)
A non-academic and readable biography of Nicholaus Steno, the seventeenth-century Dane who realised that sandstone was made of sand, applied the term 'strata' to rocks, and correctly identified fossils as the remains of actual sea shells.

Gentleman, David *David Gentleman's Coastline* (Weidenfeld & Nicholson, 1988)

Large-format picture book by one of a great generation of British illustrator-artists. Many of the pictures aren't of rocks; this serves to alert you to the charms of other stuff such as beach huts, oil refineries, lighthouses, jetties, and derelict harbours.

Osborne, Roger *The Floating Egg: Episodes in the Making of Geology* (Jonathan Cape, 1998)
A book about Yorkshire geology shouldn't be even half as entertaining as this one is, written by the volunteer curator of geology at Whitby Museum.

Phillips, John *Memoirs of William Smith, Ll. D.* (available online)
The Father of English Geology, by his nephew and assistant.

Rudwick, Martin *The Great Devonian Controversy: the shaping of scientific knowledge among gentlemanly specialists* (University of Chicago Press, 1985)
The definitive account of the 'discovery' of the Devonian, with full list of characters and extracts from their letters.

Wegener, Alfred *The Origin of Continents and Oceans* (Dover, 1966, the 1922 edition trans. John Biram)
The arguments for Continental Drift, in their well explained first form. The theory was spot on but, interestingly, Wegener got quite a bit of the detail wrong.

Welland, Michael *Sand: A Journey through Science and the Imagination* (Oxford UP, 2009)
William Blake asserted that we can 'see the World in a grain of sand'; this entertaining book, by a Cambridge and Harvard geologist, fills in the detail.

Winchester, Simon *The Map that Changed the World: The Tale of William Smith and the Birth of a Science* (Viking, 2001)
The book that's given William Smith something approaching celebrity status – which he indeed deserves. Most readers don't seem to mind the over-excited style as displayed in the title.

GEOLOGICAL GUIDES

(general, then from Yorkshire, clockwise)

General

Toghill, Peter *The Geology of Britain: an Introduction* (Airlife, 2000)
Thorough, detailed and readable, with good pictures and diagrams.

Turnbull, Ronald *Granite and Grit: a walkers' guide to the geology of British mountains* (Frances Lincoln, 2009)
The companion to this book, with much more about volcanic rocks and granite, the ancient metamorphic rocks, and the Precambrian, Cambrian, Ordovician and Silurian periods. On the other hand it has almost no fossils.

Pellant, Chris *Rocks and Minerals* (Dorling Kindersley, 1992)
Only one third of the book is rocks, with just a single picture of each. Most of the minerals in the book you'll never see in the field but they're very pretty.

LEFT Quartz liberated by the friction heat of the nearby faultline, Marloes Sands, Pembrokeshire

Crumbly Carboniferous coastline, Whitehaven, Cumbria

Walker, Cyril and Ward, David *Fossils* (Dorling Kindersley, 1992)
Handsomely produced, and wide ranging – but not, of course,
complete. Like other guides, just one picture of each species,
and the examples, from the Natural History Museum, are
fine specimens rather than the broken waterworn ones you
actually find.

Regional

Osborne, Roger and Bowden, Alistair *The Dinosaur Coast:
Yorkshire rocks, reptiles and landscape* (North York Moors National
Park, 2001)
Excellent introduction to the Cleveland Coast for the casual
visitor, but with little of the in-depth detail and occasional
silliness of Osborne's *The Floating Egg*.

Bell, Richard *Yorkshire: a journey through time* (British Geological
Survey, 1996)
A straightforward guide to Yorkshire, charmingly illustrated with
drawings and watercolours. No particular focus on the coast,
though.

ed. Brunsden, Denys *The Official Guide to the Jurassic Coast: Dorset
and Devon's world heritage coast* (Coastal Publishing, 2003)
Good illustrations, too little substance in the text, but worth the
(small) cover price for the splendidly detailed pull-out cliff section
at the back.

Edwards, Richard *Geology of the Jurassic Coast: the Red Coast
Revealed* (Coastal Publishing, 2008)
A detailed introduction to the non-Jurassic end of the Jurassic
Coast, between Exmouth and Lyme Regis.

Bates, Robin and Scolding, Bill *Beneath the Skin of the Lizard*
(Cornwall County Council, 2000)
An enlightening booklet with very useful photos.

Hardy, Peter *The Geology of Somerset* (Ex Libris Press, 1999)
Only two coastal chapters, but they're interesting coasts
(Portishead, and Triassic/Jurassic eastwards from Minehead) and
it's a well written, friendly book.

Lyle, Paul *A Geological Excursion Guide to the Causeway Coast* (NI
Environment and Heritage Service, 1996)
Lucid and enlightening text, photos and diagrams – don't visit
Antrim without it.

GLOSSARY agglomerate to xenolith

agglomerate igneous rock made of assorted lumps of pumice,
country rocks, etc, welded together with volcanic ash. It fills the
vents of extinct volcanoes.

alluvium a layer of gravel, pebbles etc left by a river or lake.

amygdale an almond-shaped white lump within a lava. It's a gas-
bubble hole, later filled with calcite mineral.

andesite igneous rock intermediate between basalt and granite: so
mid-grey in colour, moderately rich in silica, and cooling from a
mix of continental crust and ocean crust.

anticline an upward-bulging rock fold (think 'arch') as opposed to
a downward-bulging syncline (think 'sink')

ash term loosely applied to fine solid particles ejected by a volcano.
Volcanoes don't burn so this isn't ash in the normal sense.

biotite a black form of mica, commonly supplying the black specks
in granite.

bioturbation scrambled-up texture in sandstone caused by
creatures crawling through it. A useful term when you're
not sure what kind of creature (worms, shellfish) caused the
scrambling.

breccia rock containing large, broken chunks of older rock. Fault
breccia is caused by the shattering of the rocks as they move
past each other. Sedimentary breccia has not moved far before
settling down to make rock. If it had travelled, the contained
stones would be round-cornered and the rock would be
conglomerate, not breccia.

calcite the limestone mineral, assembled by living creatures out of
materials dissolved in seawater to form shell and other structures.
It can then be re-dissolved, and precipitated as mineral calcite,
rock cement, oolite, etc.

chert a concretion of mineral silica, self-assembled as the
surrounding sediments turned to sedimentary rock. The
particular chert formed in chalk is called flint.

concretion a lump, often of silica or iron oxides, formed within an
already-existing rock by chemical action. Flint is the commonest
concretion.

conglomerate sedimentary rock containing large, rounded
cobbles of older rock, separated by sand. Informally called
puddingstone, after the sort of pudding with raisins in.

cross-bedding within a sedimentary rock, a layer may have
bedding that lies on a slant. It formed from sand that was
deposited not flat and level, but on the sloping back of a
sandbank.

coprolite fossil faeces, eg from plesiosaur or dinosaur. Among
early geologists, Buckland, de la Beche, and Mary Anning all
appreciated coprolites.

diorite coarsely crystalline igneous rock that's the underground
equivalent of andesite lava. It's dark grey, sometimes with white
specks.

dolomite a form of limestone where the calcium has been
replaced by magnesium. Dolomite usually forms in super-salty
seas. It makes, and gives its name to, the Dolomite mountains
of northern Italy, where it gives splendid and very exposed
scrambling and climbing.

dolerite the intrusive form of basalt: so a dark, dense igneous
rock with crystals too small to see, forming dykes, sills and
volcanic plugs.

drift a layer of boulder clay and rubble left by a glacier: same as till.

dyke an igneous rock intruded across older beds. A dyke will
normally be roughly vertical, unless shifted by later earth
movements. Where the earth's crust is under tension, dykes
occur in 'swarms'.

fault a surface where two rock bodies have moved past each other. 'Normal fault': the upper rock body has moved downwards (the crust has stretched) – two large normal faults together form a rift valley. 'Reverse fault': the upper rock body has moved upwards (the crust has been compressed). 'Thrust fault': the upper body has moved horizontally over the top of the lower. 'Imbricate fault': several adjacent parallel faults combining into, effectively, one big one.

feldspar a silicate mineral, or rather a family of minerals, the commonest of the continental crust and one of the three constituents of granite. Feldspars have rectangular crystals, sometimes in centimetre sizes. They decompose into clay.

gabbro plutonic rock of the basalt family: so black, dense, and formed deep underground with big crystals.

goniatite seashell of spiral form, common in the Carboniferous, and direct ancestor of the ammonites.

granophyre a form of granite containing scattered large crystals – a granite porphyry.

Greensand a layer of seafloor sandstone, green when freshly broken, that lies underneath the Chalk across much of England.

greywacke a dark grey, ocean-bottom sandstone in thick layers. It's the result of underwater mudslides, 'turbidity currents'.

Great Dying the largest mass extinction, which happened at the end of the Permian period.

hornfels a featureless, brittle rock produced metamorphically from various starting rocks (eg slates, mudstones, tuff) by the heat of a nearby intrusion, such as some granite.

intrusive igneous rock that solidified below, but close to, the earth's surface: such as dykes, sills, and volcanic plugs. Intrusive rocks are crystalline, but the crystals are too small to see.

Lias A lower Jurassic sedimentary rock consisting of many alternating layers of limestone and shale. As a 'universal formation', The Lias may also refer to rocks of Lias age and position that are less than Lias-like.

Ma short for Mega annum, or millions of years ago.

Mantle the part of the earth's interior immediately below the crust. It is mechanically separate from the oceanic crust, although made of similar basalt-like material.

marmite black, viscous deposit found on breakfast tables.

microgranite a granite but with smaller crystals, not individually visible without a hand lens: the intrusive, rather than plutonic, form of granite.

mica a family of continental-rock, silicate, minerals. Because of the way their molecules are arranged, mica crystals are thin flat sheets.

muscovite a white form of mica (cf biotite, the commoner, black form).

olivine a green mineral, rich in iron and magnesium, found in basalt-family rocks and in the earth's mantle.

oolith a tiny sphere of calcite (or of ironstone), around 1mm across, precipitated out of the sea around a tiny fragment of sand or shell. Both o's should be pronounced, as in zoology.

oolite sedimentary limestone composed of calcite ooliths. In nineteenth-century usage, 'The Oolite' is a universal formation that also includes non-oolitic delta sandstones of Whitby.

ophiolite a complex of rocks from ocean floor and mantle shoved onto the surface during a continental collision: includes pillow lavas, serpentine and chert.

orthoclase a form of feldspar containing potassium rather than calcium/sodium, white or pink in colour. Orthoclase is the commonest form of potassium feldspar (K-feldspar, alkali feldspar). Pink granites are coloured by orthoclase.

pericline rocks folded into either a dome, or a bowl.

peridotite a black, dense rock mostly made of olivine, and the main constituent of the earth's mantle. Peridotite at the surface degrades to serpentinite.

plagioclase a form of feldspar containing calcium/sodium rather than potassium. It is transparent or white in colour, but sometimes pink.

plutonic an igneous rock that solidified deep underground, so with large, visible crystals. Examples are granite and gabbro.

porphyry any igneous rock that is generally fine-grained, but scattered with individual large crystals. Such rocks will have had a complex history of cooling, possibly in distinct phases.

quartz the mineral silica in its crystalline form, white or translucent and six-sided. Veins of quartz can be caused by the heat of a nearby plutonic intrusion (eg of granite), or by the friction heat of a major fault.

rhyolite volcanic rock of the sort that's high in silica. It's the same stuff as granite, but emerging as a lavaflow or small sill, rather than a huge underground lump. It's pale grey or orange-pink, and can contain scattered crystals of feldspar or black biotite.

rugose coral a thick, ridged coral, solitary rather than colonial (rugose means ridged). It and *tabulate coral* are the main corals of the Carboniferous, both becoming extinct at the end of the Permian.

serpentine a dark, red-and-black or green-and-black, mineral formed from decomposed Mantle rocks. The rock composed of serpentine is *serpentinite* but informally also 'serpentine'.

silica the commonest rock-forming mineral, whose crystals are called quartz and whose concretions are called chert (or flint, if they're in chalk).

sill an igneous rock intruded as a layer between two older beds. A sill will normally be roughly horizontal, unless shifted by earth movements.

Superposition Principle The rock underneath is older than the rock on top. The places where the principle fails (serious folding, thrust faulting, sills and intrusions) are always worth a good hard look.

syncline a downward-bulging rock fold (think 'sink') as opposed to an upward-bulging anticline (think 'arch').

tufa a spongy-looking kind of limestone, formed by calcite that has dissolved in water precipitating back out again.

tuff rock formed of volcanic ash.

turbidite sedimentary rock formed from a turbidity current (see Chapter 12). Most greywackes are turbidites, and vice versa.

unconformity a junction between two rock types where a large time-span has passed between the older and the younger. The younger rock is laid across the eroded surface of the older. Where the older rock has been tilted as well as eroded, the older and the newer beds lie at different angles – an *angular unconformity*.

Uniformity Principle The doctrine that rocks and geology form gradually, and by processes still happening in the world today, rather than by Noah's Flood, meteor collisions, and other such catastrophes. The principle was very useful in the heroic age of geology but is untrue (there have been several catastrophes, just not Noah's one).

universal formation a rock type widespread over the UK and supposed to be a single continuous layer perhaps world-wide. Examples include the Chalk, Lias, and New Red Sandstone.

vesicle in lava, a small hole formed by a gas bubble.

xenolith a foreign rock embedded inside another. In a magma such as granite, a xenolith could be a broken-off part of the magma chamber wall, or a speck of mantle material (peridotite) brought up from the depths.

THE WORLD UNFURLS

In these sketch maps, the brown areas represent continental crust: some of it may from time to time be covered in shallow seas (such as the North Sea today).

Ordovician

The continents of Laurasia (including the future Scotland and Antrim) and Avalonia (including the future England, Wales, and Irish Republic) are converging on each other across the shrinking Iapetus Ocean. Ocean sludge (turbidite) builds up in deep trenches at both edges of the ocean, but for the time being remains underwater.

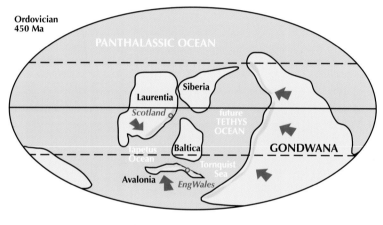

Silurian

The continents have collided, creating the Caledonian mountain range. Ocean sediments, from the Ordovician and earlier, get squashed up in the gap.

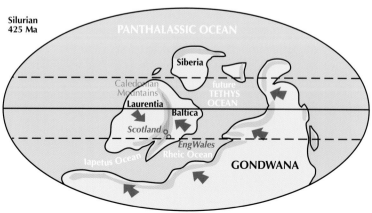

Devonian

The combined Britain lies in arid latitudes of the southern tropics. The Caledonian range is being rubbed away, creating plenty of sand and gravel for the Old Red Sandstone desert that stretches across the continent. Meanwhile, the Rheic Ocean is closing, as Gondwana approaches on a collision course from the south.

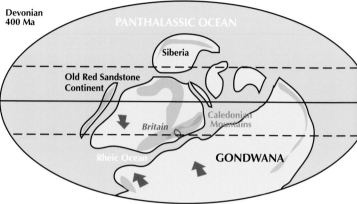

Carboniferous

Gondwana crashes the party, creating a new and huge mountain range, the Variscan (or Hyrcanian). Britain is north of the crumple zone, and gradually moving north across the Equator. Around the south pole, ice covers part of what will one day be tropical Africa as well as parts of India, Australia and Antarctica.

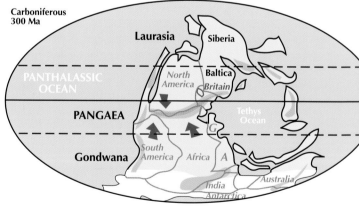

Permian and Triassic

The world's land comprises a single continent, Pangaea. In the southern temperate zone, Africa, India, Madagascar, Australia and Antarctica are all joined up. Marked in green is their shared plantlife, the Gondwana flora, named originally for part of India. Purple stars are the two locations of the freshwater fisheater Mesosaurus.

Britain lies in hot, dry northern latitudes in the middle of Pangaea. The decomposing Variscan mountains, immediately to the south, supply sand and gravel for a second red desert.

Cretaceous

Antarctica drifts southwards, and the Atlantic unzips the Pangaea supercontinent. India is just setting off in the general direction of Asia. Africa rotates north-eastwards, with the former Tethys Ocean shrinking to become the mere Mediterranean. All the other oceans are expanding at their middle ridges. The Pacific, however, has vigorous subduction zones at its east and west, so that it shrinks at the edges even faster than it expands in the middle.

Unlike the others, this map marks the shallow seas covering parts of the continental crust, as seen at the height of the Cretaceous ocean rise. The sea level is 200 m higher than now, and a shallow, stagnant chalk sea covers much of Europe.

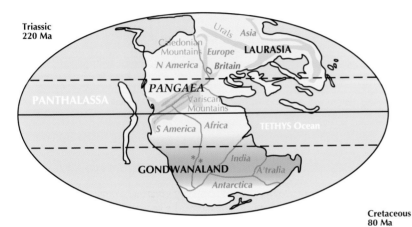

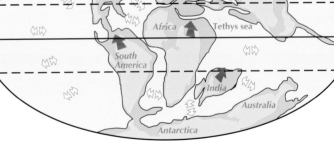

Jurassic

Britain lies in the climate zone we'd now call Mediterranean. With no land at either pole, the world is not in an ice age, and is on average 7° warmer than now.

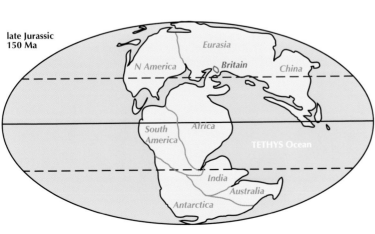

233

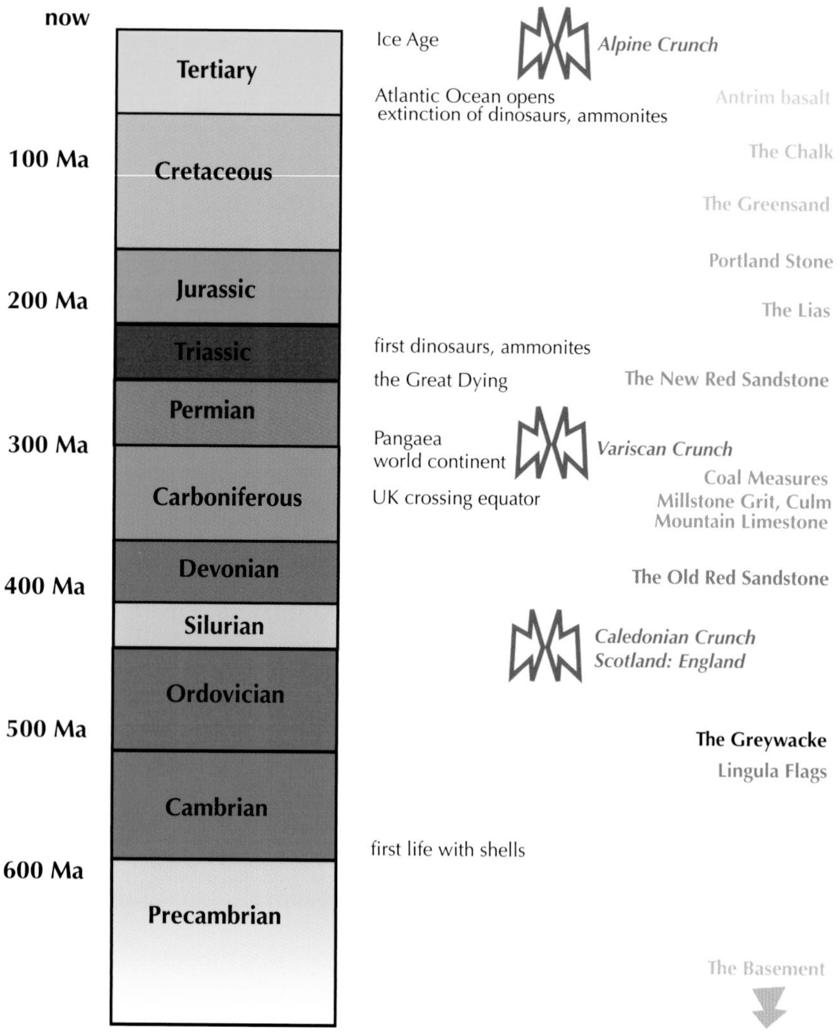

now		
	Tertiary	Ice Age
		Atlantic Ocean opens
100 Ma	**Cretaceous**	extinction of dinosaurs, ammonites
200 Ma	**Jurassic**	
	Triassic	first dinosaurs, ammonites
	Permian	the Great Dying
300 Ma		Pangaea
	Carboniferous	world continent
		UK crossing equator
400 Ma	**Devonian**	
	Silurian	
	Ordovician	
500 Ma		
	Cambrian	
600 Ma		first life with shells
	Precambrian	

Ice Age — *Alpine Crunch*

Antrim basalt

The Chalk

The Greensand

Portland Stone

The Lias

The New Red Sandstone

Variscan Crunch

Coal Measures
Millstone Grit, Culm
Mountain Limestone

The Old Red Sandstone

Caledonian Crunch
Scotland: England

The Greywacke

Lingula Flags

The Basement

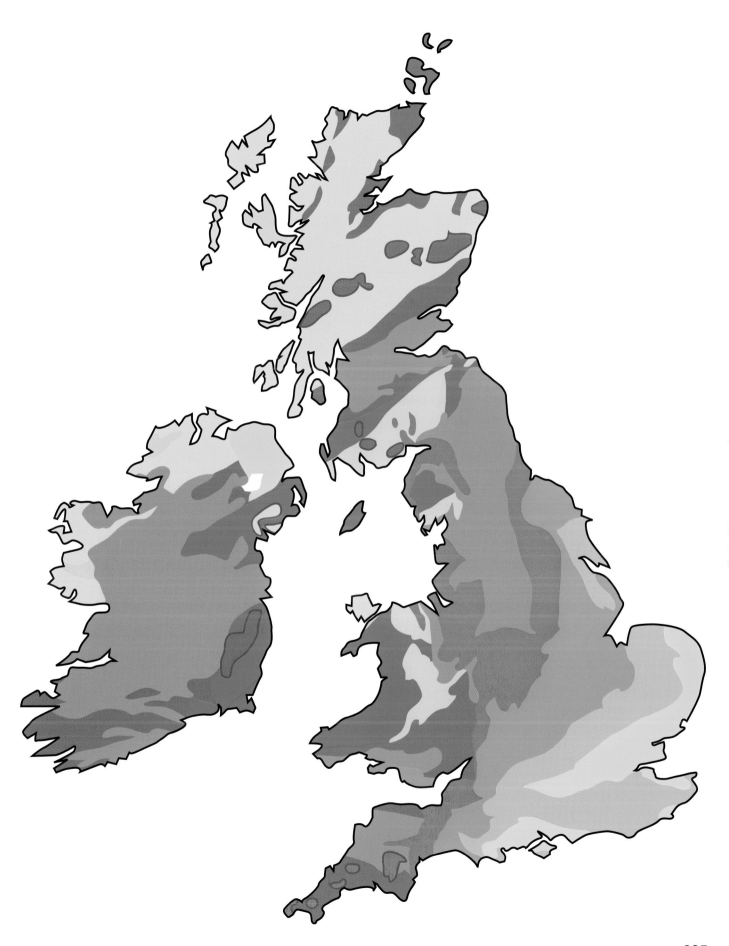

INDEX

238

ACKNOWLEDGEMENTS

Clare, thanks for your company along the coast of Yorkshire. I'm sorry you didn't get into any of the pictures, the ammonites kept sliding into the camera. Fi Martynoga suggested Fast Castle, and my genial editor Roly Smith insisted on Spurn Point (he got it in the end). Iona MacTaggart escorted me around the Lias of Glamorgan. Result, I never got to the Giant's Rock before the tide came in. I don't mind, Glamorgan was better. At Whitby Museum, Scarborough Rotunda, Portrush Tidal Zone, and the Charmouth Heritage Coast Centre, it was difficult to divide time between the absorbing exhibits and the courteous and learned curators. DG Turnbull, as well as Kate and Marek, thanks for seaside hospitality in the south. The Revd Hugh Melinsky helped me with my Latin.

Fiona Barltrop supplied fine photos of Sussex – a place which from my base in southern Scotland counts as almost in Europe. And various readers of *Granite and Grit* suggested what they would have liked but didn't get in the previous book.

I owe an obvious debt to the lucid writers listed under 'Rock Knowledge'; but in particular to Ian West, Jim Talbot and John Cosgrove for their wonderfully instructive websites, created for no payment just so that people like me could understand things. To them must be added Jimmy Wales (founder) and all the anonymous contributors to Wikipedia.

Causeway Coast, Antrim: the red laterite layer in the basalt